BIODIVERSITY

Przewalski's Horse, Edna's Trillium, The Giant Squid, and Over 1.5 Million Other Species

---

L. L. Gaddy

University Press of America,® Inc.
Lanham · Boulder · New York · Toronto · Oxford

Copyright © 2005 by
University Press of America,® Inc.
4501 Forbes Boulevard
Suite 200
Lanham, Maryland 20706
UPA Acquisitions Department (301) 459-3366

PO Box 317
Oxford
OX2 9RU, UK

Library of Congress Control Number: 2004115488
ISBN 0-7618-3089-8 (paperback : alk. ppr.)

DEDICATION

"in the prison of his days,
teach the free man how to praise…"
W. H. Auden

to

the young

L. L. Gaddy, III,
may this book help you understand and enjoy
this green and beautiful planet

and

the older

John Mark Dean, B. B. Gaddy, and Edmund Taylor, Sr.,
who have taught me and others to understand and enjoy

TABLE OF CONTENTS

Tables and Illustrations

PREFACE

I am part of all that I have met;
Yet all experience is an arch wherethro'
Gleams that untravelled world....
Tennyson, "Ulysses"

On woodland walks and on blackwater rivers, my father taught me the common trees of the southern woods. I didn't realize this knowledge had seeped into my brain until I reached college. As a graduate student, my serious study of natural history began with these familiar trees, then turned to wildflowers, and then to spiders and other invertebrates. I did my master's degree in geographical or community ecology, and my dissertation was on an ant-plant mutualism. Betwixt, between, and beyond these studies, I have cataloged the spiders of my home state, written a systematic paper on the plant genus *Hexastylis*, mapped the flora of a southern bottomland swamp, and published a guidebook to the southern Blue Ridge Mountains.

This work started out as a textbook, but, in the end, I have arrived with a group of rambling personal essays on the essential subjects that form the scientific background needed to study and understand the biodiversity of this earth. From this blend of science and personal experience, I have ended up with something more like "biodiversity-lite" than a text. It was my intention, however, to write something that would not be too serious and would be accessible to non-academics. This book investigates the diversity of species of the earth and how the study of the major subdisciplines of field biology—taxonomy, paleobiology, biogeography, evolution, population biology, ecology, and conservation biology—can help one understand these species and their diversity.

Microbiologists argue that the basic unit of life is the cell; ecologists counter by saying that the ecosystem is the most important unit. From my view as a naturalist and field biologist, the organism is what really matters in nature; the species is what biology and nature is all about. I think one must start at the basic observable unit of life and work one's way up (or down) in scale.

In the natural world, the more you know, the more there is to learn, and the more interesting things seem to become. As one observes the fascinating species on this little-known planet, the desire to communicate and discuss such wonderful things becomes inevitable. And you begin to realize that, amidst the religious, cultural, and national differences that separate our species, love for this earth and its natural beauty and bounty is the one thing that unites us all.

. . .

I would like to thank my long-time companion and friend Ms. Hu Ye, who drew Figures 1, 9, 19, and 20, and my my son, L. L. Gaddy, III, who sketched Figure 21. Dr. Lesa Dill contributed Figure 16, and Dr. John Morse of Clemson University helped me find Figure 17. Finally, I would like to thank Mr. Robin Smith of Columbia University Press who read and critiqued an early draft of this book.

INTRODUCTION

Figure 1.   Edna's trillium (*Trillium persistens*) (Hu Ye).

We all know the poster--"Love your mother." Seen from our moon, a green, blue, and brown planet with spirals and wisps of white cloud masses floats in the blackness of space, this is your earth. That green we see is a mosaic of plant life, none of which can be seen individually. The blue is Homer's "mackerel-crowded sea." The brown is the rock and soil that hold it all together. And the white is the water vapor without which we would be just another barren rock. One roughly spherical planet about 25,000 miles in circumference and one biological system with about 1.5 million named species and 3.5 billion individuals of the species *Homo sapiens*.

A spectacular array of living creatures inhabits this planet; if a man is tired of the diversity of life on earth, to paraphrase Samuel Johnson, he is tired of life itself. It seems in our solar system, it has been all or nothing. There are no other worlds that have less diversity or more diversity of life than the earth. This planet, the only one—as far as we know (and that's not really too far)—that has any life at all, much less a diversity of it. The earth has been around for about 4.6 billion years, but there has been life here for 3.5 billion years. But life here we do have. Earth's the right place for our kind of life.

Beginning with "aardvark," the first real word in the Oxford English dictionary—Africaans for "earth wolf," the grand tour of the biological diversity of our planet commences. Between aardvark and the few z-things that are known from the biosphere are everything from giant squids, which have never been seen

2

swimming in their native habitat by humans, to platypuses, a swimming mammal that lays eggs, to a few hundred thousand beetles, to a centipede-like primitive crustacean discovered in a cave in the Bahamas just a few decades ago, to a completely new phylum of invertebrates, discovered in 1995, living on the claws of lobsters in the North Sea.

Earth's biodiversity or biogeodiversity, as it might better be called, has recently become a widely discussed topic outside the circle of biologists and naturalists. There are earth watch institutes and conservation groups that give up daily updates on the biodiversity—or vanishing biodiversity—of the planet. We know the earth has millions of species of plants, animals, and other creatures, and most of us generally agree that all of these species were not created on the sixth day. The world is populated with a vast and interesting range of organisms that have adapted to particular environments. Without the myriads of environments or habitats (from the Latin "I live"), there would of course be far fewer organisms.

Ecologists often speak of coarse-filtered inventories or surveys—surveys that look at the big picture, natural communities, ecosystems, and landscape ecology versus fine-filtered inventories—works that examine genes, individuals, organisms, and species. In coarse-filtered natural history, writers look at macroscale nature and then work their way down (or never work their way down) to fine-filtered or microscale nature. To have an inkling of understanding of how the planet works, one must always be aware of what some have called "macroecology." If, however, if you want to understand what biology is really about, you must look at the species, the basic unit of nature, and the individual, the unit of evolutionary change. The naturalist, the field biologist, and the average interested observer on the street, I think, is more interested in the living thing that in the *concept* of the living thing. Interested not in the crustacean, but the blue crab (*Callinectes sapidus*) (and, yes, eating them, too); not in the Order Coleoptera (the beetles), but the powder post beetle who drills into the timbers of the house; not in invertebrates, but in the tick that carries Lyme disease; and not in mammals, but in the ever-shrinking distribution of the rare giant panda (*Ailuropoda melanoleuca*) (literally, "cat foot black and white").

Ecologists and evolutionary biologists often say that "species don't matter"; what is important, instead, in biological science are basic "laws," theories, and trends that can be seen in all species. There, of course, is truth in that approach. But species are and will always be the basic unit of nature—because we are one species, *Homo sapiens*—human beings, we must start with the species concept and work our way up to grander schemes. With an understanding of species biology and species interactions, one can only then begin to understand phylogeny, community structure, and ecosystem function and piece together the fantastic puzzle we know of as life on earth.

A young biologist once told me she wanted to study cetology, the science of marine mammals. I responded that I thought that it was a fascinating field, but whales were hard to study and even more difficult to hold. Like whales, many of the hypotheses and theories of modern biology are very interesting, but difficult to get your hands on. I propose here that we take a grand tour of the earth's biodiversity through the lenses of the various subdisciplines of biology concentrating on the organism, the species, the individual—the flesh and blood of living things.

What then do you really need to know to take such an earthly tour and to understand the biodiversity of this planet? You don't need a Ph. D. in biology or a Ph. D. in anything, for that matter. To start off, you need to know what a species is, how species are classified, where animals and plants live and why, how do animals and plant interact, and how the emergence of Man has affected the other species on the planet. Some of the basic questions we might ask are: How is this organism different from other species? What color is it, what is its morphology? Where is it found? Why is found there? What are its ecological requirements? How does it interact with other species? How long does it live? These questions have traditionally been asked by academicians from the fields of *taxonomy, biogeography, ecology, and population biology*, but these are questions any interested person would casually pose while surveying the amazing biodiversity of life on Earth. We will herein attempt to deal with some of these questions and most of these disciplines in which professional biologists study biodiversity.

In this book, I have attempted to address the major topics of biodiversity in ten loosely-connected chapters covering the major subfields that are essential to an understanding of biodiversity. Chapter I discusses taxonomy, the first (and sometimes dismal) science; Chapter II is of biodiversity and its earthly numbers; Chapter III is a brief overview of "what might have been"— paleobiodiversity; Chapter IV covers biogeography; Chapter V is about evolution (here, I am trying to cover in a few pages what Stephen Jay Gould said in 1433 pages); Chapter VI concerns population biology; Chapter VII, alternatively named "Fighting, Fornicating, and Getting Along," covers species interactions or microecology; Chapter VIII looks at the big picture, macroecology; Chapter IX is about conservation biology, rare species and environments and the preservation thereof; and Chapter X is my frail attempt to figure out where man fits into to all this.

I have selected an unlikely group of three species to help me hold together my ten essays on understanding earthly biodiversity: they are Przewalski's or the Mongolian wild horse (*Equus przewalskii* or *Equus caballus przewalskii*), Edna's or persistent trillium (*Trillium persistens*), and the giant squid (*Architeuthis dux*).

Przewalski's horse, a mammal, is the last living "true" horse. Named by the Russian explorer/naturalist, Nicholas Michailovitch Przewalski in 1879, it is native to the Altai Mountains in southwestern Mongolia. On the brink of extinction in the wild, it lives and is bred in zoos and has been recently introduced into the wild. Standing just over one meter in height, the Mongolian wild horse, as it is alternatively called, has been shown to have 66 chromosomes, compared to the 64 known in all other living species of horse.

Edna's trillium is a small, three-leaved, single-flowered herb in the lily family (Liliaceae) known from only from ravines and gorges of the Tallulah/Tugalo drainage system in the mountains of Georgia and South Carolina. It was discovered in the 1960s by Ms. Edna Garst, who explained to her husband John and professor of botany Wilbur Duncan that she could not find it in any of her wildflower books. Dr. Duncan remembered seeing something similar to the plant in a nearby gorge 20 years previously (botanists have wonderful memories).   Upon reexamination of other plants in the area, Edna's trillium was described as a new species. Because of its narrow range and low numbers, the plant is now listed as "endangered" by the Fish and Wildlife Service of the United States Department of the Interior.

The final member of our titular threesome is the giant squid. A cephalopod, the giant squid is the largest known squid and is the earth's largest known invertebrate at up to 20 meters in length with the largest eyes of any earthly creature—to 38 cm in diameter. Long thought to be a mythical sea monster, the giant squid is now known to circumnavigate the globe searching for prey (mostly other squids) or on the run from its chief predator, the sperm whale (*Physeter macrocephalus*) (a researcher from England once found over 10,000 giant squid beaks in the stomach of one sperm whale). The giant squid has been studied by man for over 100 years—recently with cameras attached to the backs of sperm whales, but no living giant squid has ever been observed or filmed in its native habitat—the oceans.

The diversity of life is an epic poem in an international language understood by all who live in the natural world and even a few urbanites who stop only every once and a while a take a look at the natural world around them.   The totality of its chapters and verses are not completely known by any one of us, but the overall theme of the work is not difficult to comprehend. We, humans, do not understand the natural flow of our birth and death, the birth and death of other species, the ecological interactions of nature, or the geological cataclysms that change and shape our earth and its life. To those of us in the industrial world, nature is an escape from the harsh realities and monotonies of work in man-made environments; to the remaining humans who still live in the forest or the wild, nature is habitat and food, the very essence of life. But to both pre-industrial and modern man, there is something primeval in the teeming life in a rich, humid swamp or an old-growth forest, and we all intuitively understand the intellectual concepts of biological richness and diversity when we walk in an alpine wildflower meadow, or swim over a coral reef with myriads of colorful fish.

As the poet e. e. cummings once said, possibly thinking of the natural world and all of its beauty and wonder, "there is a hell of a good universe next door, let's go."

Figure 2. Przewalski's horse (*Equus przewalskii*).

I

TAXONOMY:

ANIMAL, VEGETABLE, OR MINERAL? OR

DID KING PHILLIP CALL OUT FOUR

GALLANT SOLDIERS?

"What is the use of their having names," the Gnat said, "if they won't answer to them?"
"No use to *them*," said Alice, "but it's useful to the people that name them, I suppose."

Alice in Wonderland, Lewis Carroll

Is taxonomy the first science? Maybe. Adam and Eve were required to learn to differentiate the tree species of the Garden of Eden in the biblical account of the first humans. To Aristotle and the naturalists of his time, all things could be classified as animal, mineral, or vegetable. Others divided the earth, its life, and its essence into four basic elements: wind, fire, water, and earth. With such simplistic systems, early science attempted to find order in the natural world, or at least to fit the world into its own conceptual order. As man realized that some things on the face of the earth did not fit well into the ancient systems of flesh, earth, and leaf or wind, fire, water, and earth, by necessity more complex systems of classification evolved.

Long ago scientists agreed that minerals were not really a part of the classification of life (although some evolutionary biologists now think the structure of early life may have been simple clay). And, not so long ago, most taxonomists concluded that the fungi were not plants. These assumptions led to the erection of a five-kingdom system (monerans, protists, fungi, animals, and plants) of life on earth, a system that was accepted most biologists until recently. In the past few decades, however, studies of mitochondrial DNA have challenged the five-kingdom system and newer, more complex, taxonomies have emerged. Some now think that a three domain, 21-kingdom system more accurately reflects the natural divisions of the 1.5 million or so *known* species on the face of earth.

Let's go back to the beginning of the dictionary--aardvark. An aardvark is an organism, a species. Its scientific name is (*Orycteropus afer*). A flea, a humpback whale, a mahogany tree, a platypus, a human being are all species, the most basic unit of biology and of biodiversity. Species are generally defined as morphologically similar organisms or organisms that can interbreed with others of the same species. "Animal," "mollusk," "crustacean," "bug," are less basic and more indeterminate groups of creatures and taxonomic abstractions. The cell biologist may assert that the cell is the most basic unit of biology, but cells in different species of organisms may be nearly identical. The cell is one of the basic building blocks of the organism, but only the species can express the complexity of cell organization and reflect a complex system of shared taxonomic characters and lineages that may have taken eons to evolve.

The simple mnemonic devices such as "King Phillip called out four gallant soldiers" and "Kenny played chess on a fat girl's stomach," both invented to help students remember the kingdom, phylum, class, order, family, genus, and species hierarchy, might now be "*Did* King Phillip call out four gallant soldiers" or "*Did* Kenny play chess on a fat girl's stomach?" The newly-added "D" in "Did" comes from the recent addition of three "Domains"-- Archaea, Eubacteria, and Eukaryota--atop the 20 or so kingdoms. The Eukaryota are the "higher" more complex organisms, while the Archaea and the Eubacteria are primitive, microbial domains. All creatures fit somewhere, although there is usually disagreement as to where, in the three domains and multi-kingdom system.

Let's take a look Edna's trillium. Pliny the Elder or Aristotle would have easily placed it, and rightly so, in the Vegetable kingdom. Now we would say it is in the **Domain** Eukaryote, **Kingdom** Plantae, **Phylum** Tracheophyta, **Class** Angiospermae, **Order** Liliales, **Family** Lilaceae (the lily family), **Genus** *Trillium,*

Species *persistens*—common name, Edna's or persistent trillium. The accepted shorthand method of writing all of the above is the binomial *Trillium persistens* Duncan. "Duncan" signifying that the plant was first described by this botanist. If we had chosen the common dandelion (from the French "dent le lion," lion's tooth), we would have discovered that it is called *Taraxacum officinale* L. among scientists. The "L." is for Linnaeus, the Swede Carl Linnaeus, the father of modern taxonomy.

Before Linnaeus wrote *Species Plantarum* (1748) and *Systemae Naturae* (1768), compiling a list of all species known to science at his time, plants and animals were often described by long Latin sentences. For example, in 1741, in *The Natural History of Carolina, Florida, and the Bahamas*, Mark Catesby called the bullfrog *Rana maxima Americana aquatica* or the largest American aquatic *Rana*. Linnaeus shortened this epithet to *Rana catesbiana*, the scientific name still used today. To Catesby, the mockingbird was *Turdus minor cinerea albus non maculatus* (small, gray-white *Turdus* without spots), which became *Turdus polyglottos* to Linnaeus, and now *Mimus polyglottos*.

Before Linnaeus, taxonomy got even more complicated if two organisms were closely related. As attested by his writings, Mark Catesby noticed the difference between the two swamp congeners, the water tupelo (now *Nyssa aquatica* L.) and the swamp tupelo (now *Nyssa biflora* Walter). (Linnaeus actually missed the swamp tupelo, probably thinking it was the same plant as the water tupelo. The swamp tupelo was given today's accepted binominal by Thomas Walter, although Catesby was probably the first naturalist to notice the difference between the two trees.) To Catesby, the water tupelo was *Arbor in aqua nascens, foliis latis acuminatis & dentatis fructu Eleagni majore* or "tree living in water, leaves broad, flat, and acuminate & with teeth, fruit like a large silverberry," while the swamp tupelo was *Arbor in aqua nascens, foliis latis acuminatis & non dentatis fructu Eleagni minore* or just like the water tupelo but with "**no** teeth on the leaves and a fruit like the **little** silverberry." Interestingly, botanists are alone among scientists in that they still use Latin to describe new species; a new species described in English is invalid until it is described and published in Latin.

The features used to classify living things are generally referred to as "characters." Historically, most taxonomic characters were morphological features such as flower shape, pubescence, or leaf shape or pubescence in plants, while in animals, dental structure, body color, and other gross characters were used to separate species. As taxonomy evolved, however, smaller, less noticeable characters such as leaf micropubescence (in plants) or genital shape (in invertebrates) that could be detected by scanning electron microscopes became more important. In the last few decades, morphological studies of species have been supplemented (and sometimes supplanted) by karyological (chromosome) analysis, protein electrophoresis, flavonid chemotaxonomy, and DNA analysis. These analyses take place in the lab and are often extremely complex. Research is progressing at such a rapid rate, finding the biochemical "essence" of a species will undoubtedly become easier in the future.

Species that for years had been thought to be of one taxon have recently been found to be different based on electrophoretic analysis, chromosomal counts, or DNA analysis. For years I noticed a strange sedge on the rocky slopes of one of my favorite waterfalls in the southern Appalachians. Its long, blue-green leaves were unlike anything I had ever seen in the region. For over a decade, I studied this plant, its relatives, all the while thinking it was a new species. Unfortunately, the morphological differences between this plant and its closely-related congeners were just barely adequate to separate this species from others. I proposed the plant as a new species and submitted the paper to a taxonomic journal. The paper received a lukewarm reception from experts, including the botanist for whom I was naming the species. Meanwhile, another botanist had, purely by accident, collected the plant from a distant locality. He also noticed its differences from described species and subsequently conducted a chromosome count of the plant. It had a different number of chromosomes from each of its closely-related allies. Subsequently, my paper was almost immediately accepted, as it became obvious the plant was a new species.

The above story can be repeated for many other recently-described plants, as well as salamanders, insects, and other taxa. Field biologists and naturalists tend to see minute differences in species, while herbarium and museum biologists, who usually examined dried plants or animals in alcohol (so-called "alcoholic specimens"), tend to see the similarities in different species. Field biologists are generally more concerned with differences at the species level, while museum biologists are most interested in generic and familial differences and the phylogeny of various taxa. The lines of distinction between taxonomists are usually drawn at the perceptional level. There are lumpers, and there are splitters. Again, the lumpers usually work from processed specimens, while the splitters work in the field and see all the minute differences.

One taxonomist may review of genus of plants and conclude that the genus has 100 species, while another botanist may place the species number of the genus at around 35. I studied the taxonomy of the genus *Hexastylis* (Family Aristolochiaceae), herbaceous perennial plants found in the southeastern United States, for over a decade and concluded that the genus had about 12 species (Figure 3 includes distribution maps of seven of the species in this genus). A herbarium botanist, who had never seen most of the species of the genus in the wild, told me that after reviewing most known specimens of the genus (which, of course, were pressed and dried), he could not see how there were more than three species in *Hexastylis.*

A fellow biologist at a marine laboratory was a hermit crab taxonomist. Her specialty was deep-sea hermit crabs, which she rarely saw living (their collection usually involved being dredged out of the depths of the sea by research trawls and immediately dropped into alcohol). Commenting on several coastal littoral species of hermit crabs I had observed in the wild, I remarked, "Is the one with the purple hairs near the eyes a different species from the others?" She replied, "I don't know the colors of the hair, I have never seen any of them before they've been in alcohol."

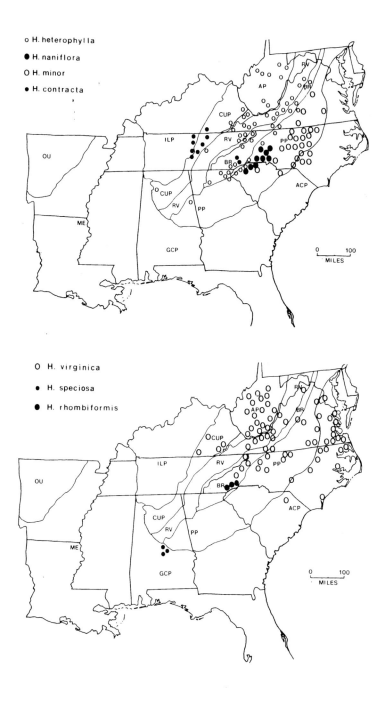

Figure 3. Distribution maps of several species of the plant
genus *Hexastylis*.

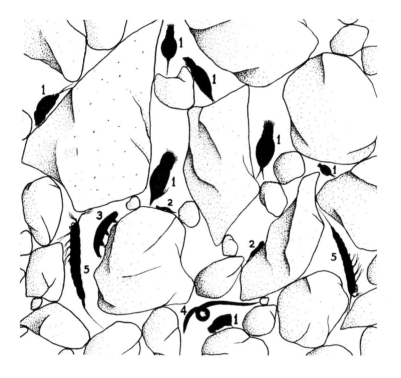

Figure 4. Animals in 1 square mm of beach sand: 1) rotifers; 2) gastrotrichs; 3) tardigrade; 4) nematode; and 5) copepod. Note size of animals compared to grains of sand (from Pennak, 1978).

II

BIODIVERSITY:

AN INORDINATE FONDNESS FOR BEETLES

…one world at a time….

last words of H. D. Thoreau

Amazingly, there is as yet no centralized computer [or other] index of [all] recorded species [the 1.5 million known species]. It says a lot about intellectual fashions, and about our values, that we have a computerized catalogue entry along with many details, for each of several million books in the Library of Congress but no such catalogue for the living species we share our world with. Such a catalogue with appropriately coded information about habitat, geographical description, and characteristic abundance of species in question (no matter how rough or impressionistic), would cost less than sequencing the human genome; I do not believe such a project is orders of magnitude less important. Without such a factual catalogue, it is hard to unravel the patterns and processes that determine the biotic diversity of our planet.

Robert May

My suspicion is that the Universe is not only queerer than we suppose, but queerer than we *can* suppose.

J. B. S. Haldane

Introduction

Many of the great books and papers in the field of natural history have asked the question: "Why are there so many species on the face of the earth?" Be it because of adaptive radiation, divine creation, or colonization from space, this planet is packed with interesting species. One of the most famous anecdotes in biology relates that a scholarly cleric (sometime told as the Archbishop of Canterbury) asks an English biologist (usually the witty J. B. S. Haldane but sometimes T. H. Huxley is named), "What can we learn about the Creator from the Creation?" Haldane replies, "An inordinate fondness for beetles!" Not knowing that there are more described beetles than any other group on earth (over 300,000 species), the cleric undoubtedly misses the point of the joke. The "Creator" was indeed extremely fond of beetles and a lot of other living things.

Species

All biological science is based on the concept of "species," distinct taxonomic units recognized by scientists. A rose is a rose is a rose; but a swamp rose (*Rosa palustris*) is not a multiflora rose (*Rosa multiflora*). Historically, recognizing species differences was based primarily on morphology. There was no way to "prove" that a species was distinct from its close relatives. Field biologists considered an organism a separate species because it was a different color or shape than another organism. If no obvious examples of hybridization existed (indicating interbreeding between the two species), they both continued to be recognized as distinct taxa. Today, biologists use chromosomal, chemotaxonomic, protein, and DNA data to determine whether a species is truly different.

The species is the most basic level of living things. The variety of species on earth is what constitutes "biodiversity." "Biodiversity" is something of a buzzword today. With good reason, however, as the mosaic and diversity of life is the living story of our planet. But just what does it mean, when one makes the statement that a place, a forest, a marsh, a city park, or whatever, is rich in biodiversity. Diversity is defined by the Oxford English Dictionary as "...different in character or quality; not alike in nature or qualities." The diversity and richness of this planet includes the giant squid, the bristlecone pine, the water strider, goatsuckers, the Venus flytrap, the large and the small, and the animal and the vegetable. And they all are united by the same basic DNA structure, the same three elements of life, Carbon, Hydrogen, and Oxygen.

The biological and geographic diversity of the earth has surpassed the wildest imagination of the ancients and of us moderns. In 1735, Linnaeus first published *Systema Naturae,* curiously the last compilation of all the species known on earth. When he published the last installation of this work in 1768, it was 2,300 pages long. Today, we have a much better idea than Linnaeus of where the earth's most diverse environments are found--tropical rainforests, coral reefs, Asian subtropical forests, the Cape of South Africa, and the Indian Ocean; however, we have no modern "system of nature," no list, no agency, no entity entrusted to keep up with the names and numbers of all the species known on the planet.

Several "all-species" or "all-biota" inventories have been proposed, but we still don't have an all-species "tabulation," a list of all species already described. The number of *known living* species on the planet is now around 1.5 to 1.7 million. Most biologists agree that there are significantly *more* undescribed species than known species. Microbiologists, in fact, estimate that as much as 99 percent of the soil microfauna consist of undescribed species. There is widespread disagreement, however, on the exact number of species that may exist on the Earth. Estimates range from three to 100 million, according with whom you are talking.

Diversity on earth is of course the product of evolution, the adaptive radiation of species. In some discussions of biodiversity, it is often forgotten that diversity is also a product of diverse environments. Early conservationists were the first to point out that it is impossible to separate a species from its habitat. Subtle differences in habitats and microhabitats produce species with minute habitat (and sometimes morphological differences). Our all-species inventories must include habitats as well as species. The ancients, with their animal, *mineral,* and vegetable, may not have been as far off the mark as we thought.

Comparative Species Richness

Some environments are obviously richer in species than others. Tropical rainforests, coral reefs, and temperate deciduous forests are generally recognized as the world's richest environments. One hectare of the richest lowland tropical rainforest in Borneo has over 1000 species of plants and animals, the majority of which are woody plants and insects. One hectare of the richest mesic forest found in China or in the southern Appalachians of North America

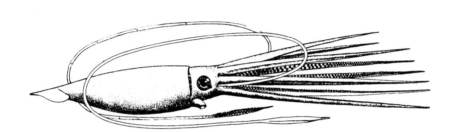

Figure 5. The giant squid (*Architeuthis dux*).

may contain over 500 species, most of which would be perennial perennial herbs. One hectare of the Great Barrier Reef may harbor over 500 species, most of which would be corals, mollusks, and fish. Finally, one hectare of dry fynbos on the Cape peninsula of South Africa may have over 1000 species of shrubs and nonwoody species. All of these environments have equally great species diversity, but they are rich in species of different taxonomic groups.

Plant Species Richness

In the vast boreal forests of Russia and Canada, the tree canopy may consist of from one to three species for miles and miles. In the dry mountains of the southwest and the west of the United States, an entire region may only have ten species of trees. In humid subtropical and temperate deciduous forests, 50 to 100 tree species may be present in a given region. These forests, however, are still depauperate when compared to the rich lowland tropical rainforest.

Lowland tropical rainforests in Borneo has been reported to have over 400 trees species. In western Amazonia, over 200 tree species have been counted in a one hectare study area. The deciduous forests of central China and the deciduous/evergreen forests of southern China are thought to have between 100 and 200 species of trees. In North America, the Great Smoky Mountains have over 100 tree species, while lowland deciduous swamp forests in the southeastern United States may have 75 species of trees (Table 1).

Table 1. Global woody plant richness:
Congaree Swamp, South Carolina and other sites.[1]

SITE	NUMBER OF SPECIES			
	TREES	SHRUBS	LIANAS	ALL
LILY CORNETT WOODS (KY)(USA)	65(3)	32(2)	14(2)	112
CHAUGA RIVER GORGE (SC)(USA)	67(7)	52(2)	13(2)	132
CONGAREE SWAMP (SC)(USA)	80(5)	50(1)	26(2)	155
GREAT SMOKY MTS (NC-TN)(USA)	123(28)	104(21)	21(12)	248
BIG THICKET (TX)(USA)	55(5)	32(2)	18(2)	105
OCALA NATIONAL FOREST (FL)(USA)	66(2)	70(2)	13	149
BLUE RIDGE PROVINCE (VA, TN, GA, NC, SC)(USA)	132(18)	174(14)	36(6)	342
THE CAROLINAS (USA)	179(17)	219(18)	45(5)	443
CAPE FLORA (SOUTH AFRICA)	NA	ca. 1000	NA	NA
PACIFIC NORTHWEST (USA)	91(11)	124(6)	5(2)	220
BORNEO	ca. 400	NA	NA	NA
WESTERN AMAZONIA	ca. 350	100-150	90-200	ca. 800

[1] Source: Gaddy (in manuscript).
"Tree" is defined here as woody plants 10 cm or greater in diameter at
1.36 meters above average ground level.
Number in parentheses indicates number of exotic species in woody
flora.
Sites listed in approximate descending order of total area. NA-not
available.

(Congaree Swamp National Park, a floodplain forest in South Carolina, at around 10000 ha in size, has about the same number of tree species—80—as the entire Pacific Northwest.)

In taxonomic terms, some genera are unusually well-represented in some regions but absent in others. The genus *Rhododendron*, woody shrubs now commonly grown as garden and yard plants in North American and Europe, has approximately 950 species worldwide. About 900 of these species, however, are found in the Himalayas of Nepal, India, Bhutan, Burma, China, and in other southeastern Asian countries. Even more amazing is the fact that more than 600 of the known species of *Rhododendron* are known from a small area of extreme northeastern India, Myanmar (Burma), and extreme northwestern China.

Other examples of rich concentrations of species of one genus are found in the plant genus *Carex* (Cyperaceae)—over 500 species in North America and an undetermined number of species in China. And Hawaiian entomologists report that there may be over 300 species of the fly genus *Drosophila* in the Hawaiian Islands.

Invertebrate Species Richness

There are over 300,000 species of the Order Coleoptera, the beetles. Beetles, at least in species richness, outnumber all other species of life. The beetle fauna of the lowland tropical rainforest is indeed "inordinately" rich, as Haldane described it. Both Darwin and Alfred Russell Wallace, who spent a good bit more time in the field than did Darwin, were fascinated by the species diversity of beetles. Wallace, in his classic, neglected work, *The Malay Archipelago*, which was dedicated to his friend and colleague Charles Darwin, discussed his travels and collections (he collected over 80,000 specimens of plants and animals) of beetles in central Borneo:

> When I arrived..., I had collected in the four preceding months, 320 different kinds of beetles. In less than a fortnight, I had doubled this number, an average of about twenty-four new species everyday. On one day, I collected sixty-seven different kinds, of which thirty-four were new to me...so that I obtained altogether in Borneo about two thousand different kinds [of beetles] of which all but about a hundred were collected at this place [an opening in an extensive lowland tropical rainforest], and on scarcely more than a square mile of ground.
>
> A. R. Wallace, *The Malay Archipelago*

H. W. Bates, famous for his studies of mimicry in butterflies, collected over 15,000 species (mostly invertebrates) from Amazonia in the mid-1800s (Wallace actually accompanied him on some of his collecting trips); *8,000* of these species were *new* to science.

In 1980, Terry Erwin, a tropical biologist, found over 1100 species of beetles in the canopy of one tree, *Luehea seemannii* (Leguminosae) in a tropical rainforest in Panama. E. O. Wilson's reported 43 ant species of 26 genera from *one* tree in Tambopata Reserve in Amazonian Peru. 172 species of ants were found in one square mile of lowland rainforest in New Guinea; 219 species in one square mile of a cocoa plantation/forest complex in Ghana;

272 species/square mile in a lowland tropical rainforest in Brazil; and, 350 species/square mile in a lowland rainforest in the western Amazon. Finally, in less than one square meter of soil in an Australian sample, a graduate student confirmed that there were 172 species of mites.

Figure 4 is an illustration of the microfauna of a just over one square millimeter of a sandy beach. Note that the groups pictured—rotifers, tardigrades, etc., are not different species of a single genus or family, but are different species of different phyla. It has been estimated that only one percent of all the species of the earth's soil microfauna has been described.

The Great Barrier Reef has over 4000 species of mollusks. In freshwater mollusks, species richness peaks in some drainages where numerous endemic species have arisen. The Tennessee River drainage in North America had around 200 species of mollusks at one time. Due to impoundment of the river by an extensive system of dams, only about 100 species are now known from the river and its tributaries.

Fish Species Richness

The Indian Ocean and the south Indo-Pacific region far outrival all other areas of the world in marine fish species with over 3000 species. The Great Barrier Reef of Australia is thought to have around 1500 species of fish. In freshwater environments, the Amazon River basin's fish fauna, over 2000 species, far outnumbers its closest rival. This is quite astounding compared to the 375 species known from the Mississippi drainage in North America and the species richness of other major river systems of the world. In Lake Malawi in western Africa, nearly all of the over 200 species found there are endemics.

Individuals

The rich lowland tropical rainforest, without doubt, harbors a most impressive array of plant and animal species. There are, however, relatively few *individuals* of each species. Compare the lowland tropical rainforest where one must walk a considerable distance before encountering two trees of the same species to the boreal forest where *all* trees may be the same species for miles on end.

Great numbers of individuals of one species do not usually occur in tropical trees or in beetles, but in other organisms in other places, such as the high latitudes of the temperate zone. During the voyage of the *Beagle*, Darwin spoke of the abundance of krill *(Euphasia superba)* in the ocean near Tierra del Fuego in South America:

> In the sea around Tierra del Fuego, and at no great distance from land, I have seen narrow lines of water of bright red colour, from the number of crustacea, which somewhat resemble in form large prawns. The sealers call them whale food. Whether whales feed on them, I know not; but terns, cormorants, and immense herds of great unwieldy seals, on some parts of the coast, derive their chief sustenance from these swimming crabs.
>
> "The Voyage of the Beagle," Charles Darwin

These "swimming crabs" are minutely small shrimp. And, we now know that whales *do* feed on the krill. At around 60 degrees latitude north and south, krill still colors the waters of our cold seas a deep pink.

Primitive protozoans, dinoflagellates of the genus *Gonyalaux*, also occur in vast numbers periodically in the oceans. Giants "blooms" of these creatures are referred to as the "red tide." They are well-known because they saturate the sea water in which they occur with concentrations of toxic metabolites that can kill thousands or millions of individuals of other creatures such as fish, crabs, and marine life. The red tide, red by day and bioluminescent at night, has been reported to reach densities of *20-40 million individuals per cubic cm* of seawater. In other words, in such red tides, there are more *Gonyalaux* in 300 cubic cm of seawater (imagine a vase of seawater with a square base 10 by 10 cm wide and sides 30 cm tall) than there are individuals of the species *Homo sapiens* on the entire planet Earth. And file the following under "global biogeochemical interdependence": a climatic researcher recently reported that when winds blow mineral rich sands from the Sahara to the Caribbean, large "blooms" of the red tide may occur and can be seen from *space*. And others have reported massive, toxic dinoflagellate outbursts induced by pollutants in coastal waters in the United States (from large hog farms) and in developing countries.

Among other reported concentrations of millions of individuals of a given species is the march of the Christmas Island red crabs (*Geocarcoidea natalis*) on the remote Christmas Island in the southern Indian Ocean. Here, during mating season, millions of the endemic crabs move from the forests of the island overland toward the sea where they mate and lay eggs. They move in such masses—it is estimated that there are 100 million red crabs on the island—that human life on Christmas Island is brought to a standstill for a short period of time. And further in the annals of world's largest invertebrate concentrations are the 17-year cicadas (*Magicicada* spp.). Some local colonies of Brood X (hatching in 2004) were reported to have produced *trillions* of individuals.

Ants are among the few organisms that are rich in species and individuals. It is often said anecdotally (how could one prove the statement anyway?) that the weight of ants is greater than that of all other animals on the face of the Earth. In Amazonia, it is thought that one-third of the entire animal biomass of the rainforest is ants and termites, and that each hectare of soil there contains 8 million ants and 1 million termites. And along the Ivory Coast of Africa, in savannahs there may be up to 20 million ants per hectare. When it comes to densely-packed colonies, ants rival the dinoflagellates mentioned above and also surpass our own prolific species. A supercolony of *Formica yessensis* on the coast of Hokkaido, Japan's northernmost island, was reportedly composed of 306 million workers and 1,080,000 queens living in 45,000 interconnected nests. That ant city is over 10 times larger than the largest concentration of *Homo sapiens*. There are recent reports of a massive colony of the globe-trotting Argentine ant (*Linepithema humile*) that supposedly extends from Italy, where the ant was introduced (only 80 years ago), to Portugal, a distance of 6000 km. If this colony turns out to be one genetic mega-individual, it may rival the Great Barrier Reef as the largest living structure on the face of the earth. The Reef is around 2000 km long and contains billions of individual coral polyps.

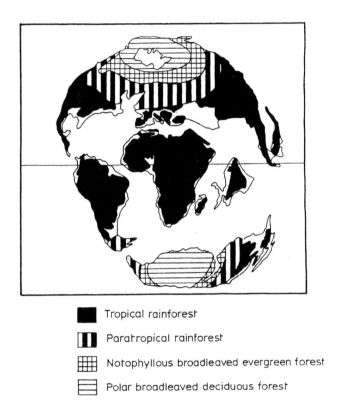

Tropical rainforest

Paratropical rainforest

Notophyllous broadleaved evergreen forest

Polar broadleaved deciduous forest

Figure 6. Eocene map of the western hemisphere
(from Brown and Gibson, 1983).

III

PALEOBIOLOGY:

LARGE, WINGED CREATURES,

SEAMONSTERS, AND DINOSAURS

No one knows how many species have existed since the beginning of the earth, but we do know that millions of species and entire faunas and floras have become extinct in the over 500 million years since primitive life began on earth. Furthermore, we know from the fossil record that many of the earthly creatures, plants, and environments of the past were very different from what we now see on the planet. Just as speciation is a part of evolution, so is the extinction of species.

Extinction is an integral part of the process of natural selection and speciation, the process driving life to diversify and embody itself in the myriad of forms we see on this planet today. Mathematical biologists have calculated that just as there are speciation rates, there are rates of extinction. Extinction has proceeded, at times, in what appears to be orderly natural rates, and has also been inexplicably sudden. Most of these sudden extinctions involved large numbers of species. Most paleontologists now agree that there have been at least five major mass extinctions in the history of life on earth (Table 2).

Death of the Dinosaurs: the Last Major Mass Extinction

No living human or living anything has ever encountered a dinosaur, but the word and the fanciful creatures that this term immediately brings to mind is known by nearly every educated human on the earth, including many still in kindergarten. The "dino"saurs were, of course, the "terrible" lizards that roamed the earth from the Permian through the Cretaceous periods of the late Paleozoic and Mesozoic Eras   (see Table 2). Every child who has

Table 2. Geologic time and paleobiodiversity.

ERA	PERIOD	EPOCH	MYA	ANIMALS	PLANTS
CENOZOIC	QUATERNARY	Holocene	0.01	Rise of the Mammal	Angiosperms
		Pleistocene	2		
	TERTIARY	Pliocene	5		
		Miocene	23		
		Oligocene	34		
		Eocene	55		
		Paleocene	65		
MASS EXTINCTION					
MESOZOIC	CRETACEOUS		141	Dino-saurs	Early angiosperms
	JURASSIC		202		
MASS EXTINCTION					
	TRIASSIC		245	Rise of the Dino-saurs	Gymnosperms, giant horsetails, lycopsid forests
MASS EXTINCTION					
PALEOZOIC	PERMIAN		290	Birth of Dino-saurs	Ferns, seed ferns, tree ferns, lycopsid (club-moss like) forests, early gymnosperm (conifer) forests
	LATE CARBONIFEROUS		323		
	EARLY CARBONIFEROUS		363	Early amphib-ians	
MASS EXTINCTION					
	DEVONIAN		409	Primi-tive fish, amphibians, and insects	Forests
	SILURIAN		439		Rise of land plants
MASS EXTINCTION					
	ORDOVICIAN		510	Tril-obites, marine life	First land plants
	CAMBRIAN		570		Marine flora
PRE-CAMBRIAN			4500	Primi-tive life forms only	Primitive life forms only

ever been through the American education system since 1950 has been bombarded with the knowledge of dinosaurs. In fact, most kids, including my own, probably know more about dinosaurs, their habitats, their generic names, and their ferocity, than they do about living creatures. Many writers and editors of popular scientific books and periodicals continue to say that people have difficulty with Latin names and, for that reason, often leave them out of publications. The study of dinosaurs by children has proven this theory to be complete hogwash. My son knew at least 25 Latin names of genera (e. g., *Tyrannosaurus, Stegosaurus*, etc.) of dinosaurs before the age of ten and can now remember over half of them.

With the exception of a few publications, including the wonderful *Dinotopia* series, dinosaurs are depicted as fierce creatures, but, in truth, they were complex, highly-evolved, beautiful reptiles. *Parasauropholus* and the hadrosaurs, for example, were large vegetarians with strange crests that may have been used for producing warning or communicative sounds. And behold the wonderful *Stegosaurus* and its close relative *Kentrosaurus*, both with eerie protective back plates and defensive spines. One could spend a lifetime studying dinosaurs, and many have done so.

Dinosaurs suddenly disappeared from the fossil record near the end of the Cretaceous period. The theory that the extinction of the dinosaurs was facilitated or even brought about by a sudden change in climate caused by a large (10-20 km wide) asteroid (comet, meteor ?) crashing into the earth just off the Yucatan peninsula in Mexico is now widely accepted. This asteroid's impact may have created a nuclear winter and altered the global climate for nearly a decade, creating a cooler, cloudier climate.

How do we know dinosaurs existed? We have an extensive fossil record from nearly all over the earth. There are footprints in North America, fossilized eggs in Mongolia and China, and bones and teeth on nearly every continent. Unfortunately, the fossil record is incomplete and extensive speculation is often involved when dinosaur behavior is discussed (especially in movies). The record does exist, however, and is relatively continuous until the late Cretaceous.

Other Mass Extinctions

As may be seen in Table 3, four other major mass extinctions have occurred, according to the fossil record. Most of the extinctions seem to be results of the earth colliding with foreign bodies such as comets, meteors, or asteroids. In the early Paleozoic, a mass extinction wiped out approximately 57% of the existing genera of primitive mollusks and annihilated many species of primitive trilobites. At the end of the Devonian Period, another extinction destroyed many ammonoides and nautiloids (mollusks related to the present-day *Nautilus*) and wiped out the most common gymnosperm of early primitive forests—trees of the genus *Archaeopteris*.

The most devastating mass extinction occurred between the end of the Paleozoic and the beginning of the Cenozoic, at the end of the Permian Period. Here, it has been estimated that 75-96% of all genera became extinct. The last trilobites were completed eliminated; the number of marine species may have been reduced from 250,000 to a mere 10, 000. After the Permian extinction, there was a rapid radiation of quillworts or horsetails, which were giants compared to those known today. Forests with giant clubmoss-like plants (lycopsids) also became more abundant in the Triassic Period, according the fossil record, along with new and varied species of conifers (gymnosperms). This extinction was followed, about 50 million years later, by another mass extinction at the end of the Triassic Period that dramatically reduced the number of ammonoids that existed at the time.

The Cretaceous-Tertiary extinction, the last major mass extinction in the fossil record, shows up dramatically in the fossil record. It marked the end of the dinosaurs, but it allowed for the rise of mammals. Some geologists point out that the beauty of extinction is to make way for the coming of newer forms of life. Without the last major extinction, there would have without doubt been no *Homo sapiens*. The lines of this chapter may have instead been written by a descendant of a velociraptor.

Other Paleofloras and Paleofaunas

Pleistocene Mammals

A varied fauna of large mammals was found in North America from the Hudson Bay to Mexico during the Pleistocene. This large fauna included massive creatures like the gomphotheres, who purportedly ate from the limbs of large-fruited tropical trees. Saber-toothed cats lived in the suburbs of Los Angeles, or at least what was to become L. A. suburbs; mammoths roamed the prairies of the Midwest (it has been recently proposed that Indian elephants, not bison, hould be reintroduced into the North American prairies); and gigantic beavers were found on the rivers of the west.

But around 10,000 years ago, near the end of the Pleistocene epoch (Table 2), many of the species of this fauna disappeared. Most vanished species, we think, become extinct through natural processes; but, in the case of the large Pleistocene mammals, it has been proposed that this extinction was caused by man. The end of the large Pleistocene mammals seems to coincide with the immigration of Asian man to the North American continent via the Bering land bridge. One theory points out that the mammals had no predators that were bilaterally symmetrical and bipedal in motion. It would have been easy, therefore, for man to sneak up on creatures unfamiliar with our body shape and hunting style. Indeed, a similar event happened in the 1800s on the island of Mauritius—the arrival of man—that led to the extinction of another famous creature, the dodo (see Chapter X, Abundance Challenged).

Seamonsters and Large, Winged Creatures

The gigantic sea reptile *Kronosaurus queenslandicus* swam in the prehistoric inland seas of Australia. It was about half the size of a large whale and was one of the fiercest predators ever known. One paleontologist noted that *Kronosaurus* could have "eaten *Tyrannosaurus rex* for breakfast." The seas were also filled with predatorial reptiles such as the Loch Ness monster-like *Elasmosaurus* and the "fish-lizard" *Ichthyosaurus*. The species-rich Indian Ocean yielded large fish similar to the coelacanth (*Latimeria chalumnae)*, the "living fossil" rediscovered off the coast of South Africa in this century.

The pterosaurs were flying reptiles, some carrion eaters, some fish eaters, and some plant eaters. Among the pterosaurs number two species that vouch for the distinction of being the largest winged creature ever to exist. The quetzalcoatlus, notably *Quetzalcoatlus northropii*, one of these giant fliers had the wingspan of a small propeller-driven plane. This beast was four to five times larger than the largest bird known today, the giant albatross (*Diomedea exulans*). And, speaking of wingspans, fossilized Carboniferous dragonflies have been found with wingspans up to 63 cm (around two feet).

Prehistoric Plants

Giant predatorial animals are, of course, more striking than prehistoric plants, but the extinct flora of this planet's history is worth noting here. Land plants appear to have arisen in the Silurian Period of the Paleozoic Era, over 430 million years ago. Early land plants were similar to our present-day clubmosses, ferns, and gymnosperms (pines, cypresses, and other cone-bearing trees). In the Devonian Period, about 400 million years ago, the first forests, dominated by the gymnosperm genus *Archaeopteris* were found in floodplains. *Archaeopteris*, a precursor to the gymnosperms, had a trunk that was up two one meter in diameter. Gymnosperms, ferns, tree ferns, and forests of the clubmoss-like lycopsids dominated the earth during the Carboniferous period. In the Triassic gingko-like conifers, tree ferns, lycopsid forests, and giant horsetails were the dominant plant species. Finally, only in the late Mesozoic Era did the first angiosperms or flowering plants arise. Gingkos and gingko-like species were widely distributed in Asia, however, as late as the Tertiary period. *Metasequoia*, the dawn redwood, and other needle-leaved deciduous gymnosperms [related to bald cypresses (*Taxodium* spp.)] were widespread in several Cenozoic floras. Interestingly, both gingko and the dawn redwood, which was only discovered to be extant 40 or so years ago, are widely planted today, but are both practically extinct in the wild in China where they are native. Finally, in the late Cenozoic Era, angiosperms diversified, radiated, and achieved the full glory we now know.

That the earth has changed so dramatically spatially and temporally can no better be seen than in an amazing excavation in Antarctica. This dig revealed that during the Cretaceous, there were complex polar forests in Antarctica. There were floodplain communities of taxodiod conifers [related to the bald cypress (*Taxodium*) of southeastern North America], along with conifer-fern-angiosperm thickets along rivers and streams. Araucarian (a group now restricted to southwestern South America) conifer forests and gingko thickets were also part of this flora.

To understand the global plant geography of today, it is important to know and understand some of the elements of the past floras and paleoclimates of our planet. It helps, for example, to be aware of the fact that there once was a deciduous arboreal flora that extended across the entire northern hemisphere. Such information makes senses when one analyzes the current biogeography of the genus *Quercus* (the oaks) or the genus *Acer* (the maples), both of which are found in Europe, eastern Asia, and eastern North America.

Paleobiogeography: Drifting Continents

Alfred Wegener, a German meteorologist, proposed the idea of continental drift, or moving continents in an early 20th century paper. His idea was, for a time, to be a subject of jokes in geologically circles, but the final joke was on the close-minded geologists. The idea that the geography of the earth has not always been like it is today, now termed "plate tectonics," is probably the most revolutionary in the history of geology.

There is good evidence that the earth consisted of two giant supercontinents—Gondwanaland and Laurasia—about 200 million years ago. South America, Australia, Africa, Antartica, India, and portions of the east coast of North America were part of Gondwanaland. Asia, Europe, and the remainder of North America came from Laurasia. Some geologists assert that at one time in the history of the earth, all of the known continents were fused together as part of the continent Pangaea (meaning "all lands" in Greek).

Plate tectonics theory asserts that the continents are crustal plates floating on a fluid-like base under the earth's crust. Since the beginning of the earth, these plates and their continents have been floating around on the surface of the globe, like ships at sea. Periodic collisions of the continents have resulted in the large mountains of the earth—the Appalachians were formed when Africa collided with North America, and the Himalaya was formed when India collided with Asia. Today, it is generally thought that there are over 20 plates making up the surface of the earth, some overriding others, some disappearing into trenches and becoming smaller, and some plates, in zones of spreading sea floors, becoming larger.

The importance of plate tectonics is biogeography and paleobiogeography lies in the fact that the distribution and connections of former continents often help explain the present distribution of plant and animal species. For example, many plant genera and plant families previously found on Laurasia only exist now in North America, Europe, or Asia. The southern beech, *Nothofagus*, only occurs in Chile, New Zealand, and Australia, all formerly of Gondwanaland. And marsupials, pouched mammals now principally found in Australia and South America [*Didelphis virginiensis*--the Virginia opossum—is a North American exception], probably originated on Gondwanaland.

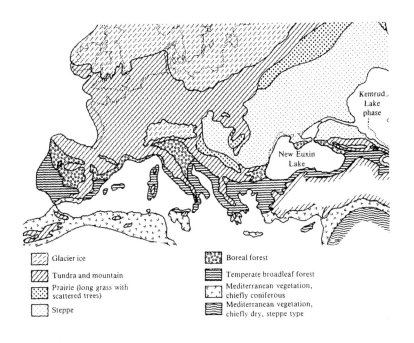

Figure 7. Europe in the Pleistocene 18000 years ago: note extent
of ice and boreal forest (from Brown and Gibson, 1983).

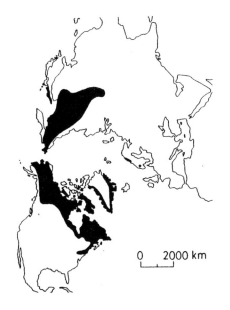

Figure 8. "Circumboreal" distribution of *Rhododendron lapponicum* (Ericacea), a dwarf azalea.

IV

BIOGEOGRAPHY:

POLAR BEARS AND PENGUINS

Distribution and Migration of Species

I was giving a lecture on introductory biogeography one day, befuddled by my students' lack of knowledge of basic global distributions, when a group of 2nd graders passed by in the hall. I grabbed an intelligent-looking girl and asked her, in front of my class of 18 and 19 year-olds, "What's the difference between the north pole and the south pole?" In one breath, she sorted out what most of the older students did not know (or did not know they knew): "The North Pole has polar bears and Santa Claus, the South Pole only has penguins."

Biogeography attempts to answer the questions of where living things grow or live and why. Organisms are not randomly distributed across the face of the earth. The geographical distribution and microhabitats of a species, for example, tell us a lot about that species' preferences in habitat or historical distribution. Some plants grow only on limestone-derived soils, some are found above 7000 meters in elevation, and others may occur in association with other plants. Some animals are found with only plants on which they feed. The pipevine swallowtail (*Battus philenor*) lays eggs on just two or three species of pipevines (*Aristolochia* spp.). If captured and offered other plants of the same family, the female pipevine will not lay eggs. Many parasitic (solitary) wasps collect, paralyze, and oviposit on one species of fly, grasshopper, or other insect. Some beetles are found only in the brood chambers of ant nests. The spores of some ferns, fungi, and bacteria, on the other hand, are wind dispersed, accounting for the fact that many species of fungi are found around the world. The credo of mycologists and microbiologists is: "everything is everywhere; the habitat limits."

Other distributions are controlled more by historical origin and evolutionary history.   Penguins evolved on the southern supercontinent Gondwanaland (Antarctica, Australia, etc.) (Chapter III) and, being heat-intolerant creatures, have never been able to expand their ranges across the tropics into the northern hemisphere (they have made it to the equatorial Galapagos Islands). Polar bears, on the other hand, evolved in Laurasia, the northern protocontinent and have never made it across the foreboding (at least for them) tropics (some bears, however, have crossed the Panamanian land bridge and have colonized South America).   On a drastically reduced scale, a study of six parallel ravines on the Blue Ridge front of the southern Appalachian Mountains in the United State, each less than one mile apart and all with nearly the same microecology and soil type, revealed that the spring ephermeral herb trout lily (*Erythronium americanum*) occurred in two of these ravines and was completely absent in the other four. Here, there, like the distribution of the polar bears and penguins, historical accident has played a strong role in distributions—the seeds of trout lily just happen to be brought into these two ravines by ants or birds and did not make it into the other ravines.

To get a better understanding of what one can learn from the study of distributions, consider the following species and their distributions.

Oconee bells (*Shortia galacifolia*) (Diapensiaceae). This striking plant species (Figure 9) is a member of a genus found only in Japan, Taiwan, and a few counties in eastern North America in North Carolina and South Carolina. This genus was possibly

Figure 9. Oconee bells (*Shortia galacifolia*) (Hu Ye).

widespread in the Arcto-Tertiary flora that covered much of the northern hemisphere's temperate zone over 30,000 years ago.

Dwarf-flowered heartleaf (*Hexastylis naniflora*) (Aristolochiaceae). This plant is one of 12 or so species of a genus of perennial herbs found in the southeastern United States. This species is found only in North and South Carolina and is even further limited to acidic, rocky and sandy soils in the middle of the clay-dominated Carolina Piedmont. It is locally abundant where it is found, but its total geographic range is miniscule. Biogeographers call such species "narrow endemics." See Chapter VI for a discussion of the history and distribution of Edna's trillium (*Trillium persistens*), another "narrow endemic."

Least trillium (*Trillium pusillum*) (Liliaceae). Least trillium is found throughout the southeastern United States. Its populations are widely separated. Some populations are 300-400 miles from the nearest other population. Some taxonomists have speculated that the complex is made up of several varieties, with little or no gene flow between the varieties. Others speculate that DNA analysis will eventually show that all of the varieties are separate species.

Yucatan cerripede (*Spelenectes tulumensis*). This weird, swimming centipede-like crustacean is known only from a few water-filled caves (associated with *cenotes*, open water springs and sinkholes that connect to caves) near Tulum in the Yucatan peninsula of Mexico. Other members of this genus are found in the Bahamas and in other distant cave localities. The biogeography of cave organisms is quite strange. Because of the isolated nature of caves, species in a single genus of cave beetles, for example, will be found hundreds of miles apart.

West Virginia white (*Pieris virginiensis*). The West Virginia white is closely related to the European cabbage white (*Pieris napi*) and, like the cabbage white, feeds on plants in the cress family (Cruciferae). The West Virginia white is found in woodlands in eastern North America where it feeds only on plants in the cruciferous genus *Cardamine*. Because this *Pieris* does not fly long distances between fragmented blocks of deciduous woodlands, it has become rare and local in the southern Appalachians.

Single-sorus spleenwort (*Asplenium monanthes).* This spleenwort is found all around the globe. Originally described from South Africa by Linnaeus, it is found in the Neotropics, the Palaeotropics, and in the Palearctic of North America as a disjunct. It is intolerant of severe frosts and drought; therefore, it is generally found in wet gorges at lower elevations in the temperate zone and in foggy mountains at higher elevations in the tropics.

The giant squid (*Architeuthis dux*). The giant squid has been collected around the world. Though little is known of its microhabitats and habits in the oceans, it is known that the giant invertebrate prefers cooler waters. It has washed up on the shores of Newfoundland, the beaches of New Zealand, has been collected off the coast of Japan, and is known from Arctic and Antarctic waters. It has been suggested that the giant squid may actually swim under the North Pole during its movements around the globe. One writer, commenting on the powerful swimming capabilities of the large squid, speculated that the squid could probably get from northern polar waters to southern polar waters in less than a month.

Distribution by Migration

Birds, of course, are most famous for long-distance migrations. The Arctic Tern (*Sterna paradisaea*) flies the length of the globe annually, wintering on subantarctic and summering near the edge of the Arctic Ocean. Many of the passerine or perching birds from eastern North America and Europe spend winters in the tropics where insects are available. Warblers and ducks are famous for their night migrations. Both of these bird groups are thought to navigate at night by using star and constellation positions. Butterflies also travel far on the wing. The monarch (*Danaus plexippus*) winters along the south-central California coast and in the mountains of central Mexico. Some monarchs fly all the way from Canada to Mexico, with the help of cold fronts, in about a week.

The sockeye salmon (*Oncorhynchus nerka*) historically was known to migrate up the Salmon River from the mouth of the Columbia River to Redfish Lake in central Idaho to breed in fresh water. This journey, which thousands of the reddened (hence, "Redfish" Lake) salmon used to make before dams were built on the Columbia and Salmon rivers, was over 950 miles long. Today, fewer than 50 fish annually make it to Redfish Lake. The sockeye and other salmon can smell the waters in which they hatched and will not end their migration until they return to their natal waters.

The *catadromous* American eel (*Anguilla rostrata*) lives a migratory life nearly opposite to that of the sockeye salmon, an *anadromous* species. The young migrates to fresh water where they grow to adulthood and then return to salt water to breed. Adult eels from the rivers of the eastern seaboard of the United States migrate annually into the south Atlantic headed for an area between the

Equator and the Tropic of Cancer where they encounter a rich tropical sea of floating seaweed and associated invertebrates—the Sargasso Sea. Here, the American eels, along with the European counterparts, who travel farther than the American eels do, breed each year.

There are only seven or so known species of marine turtles, but they have circled the globe, migrating through most of the tropical world. The leatherback (*Dermachelys coriacea*) has been captured in the Arctic Ocean *north* of the Bering Strait, giving it the possible distinction of the most cold-tolerant of all reptiles [along with the garter snake (*Thamnophis sirtalis*), which ranges into north Canada, and the northern adder (*Vipera berus*), which has also been found north of the Arctic Circle (in Russia)]. The seas are also filled with migrating great whales, some of whom swim from arctic waters to tropical waters on an annual basis. The sperm whale (*Physeter macrocephalus*), in fact, may swim the entire length and breadth of the oceans in search of its prey the giant squid (see above). Whalers learned these routes early and, as in *Moby Dick*, often tracked individual pods of whales for thousands of miles. Today, boats devoted to "whale watching," and even "whale interaction," in Maine, California, and the Gulf of California, depend on the migratory periodicity of many species of whales to keep their clients happy.

Faunistic and Floristic Regions

Wallace's early map (1876) of the faunistic regions or "provinces" of the world still stands as an accurate piece of biogeographic work. Biologists, biogeographers, and taxonomists still use Wallace's terms for the world's six zoogeographic regions: "Nearctic" (North American region), "Palearctic" (Europe, north and central Asia, and north Africa), "Ethiopian" (Africa), "Oriental" (India, south and southeastern Asia south to Borneo), "Australian," and "Neotropical."

Tahktajan (1986) has compiled general and detailed maps of the floristic regions of the world (Figure 11). These regions are based on taxonomic similarities in plant species and not structural similarities in vegetation. In his map, there are six floristic "kingdoms," 35 floristic "regions," and around 150 floristic "provinces." The names of the floristic kingdoms of Tahktajan are well-known: the "Holarctic" (sometimes divided into the Nearctic, the New World temperate, and the Holarctic, Old World Temperate), the "Palaeoantarctic," the "Neotropical," the "Palaeotropical," the "Australian," and the "Cape" kingdoms. With nine regions, the Holarctic is the most widespread floristic kingdom, ranging throughout the temperate zones of North America, Europe, and Asia. The Cape Kingdom is the smallest kingdom, being found only in the Cape of Good Hope area at the southern tip of Africa. For such a small area, the Cape flora is extremely rich in species. Over 8,500 species have been reported from this kingdom. Eight plant *families* and over 6,200 species of the Cape flora are found nowhere else in the world. The "fynbos" (a type of shrubland comparable to the "chaparral" of other mediterranean-type vegetation zones) of the Cape kingdom may have as many as 2,000 species of shrubs.

Biologists and biogeographers often refer to both vegetation and floristic regions in the same sentence to further pinpoint where a plant or animal is found. For example, a species may be said to occur in "neotropical" not "palaeotropical" rainforests. Or vegetation and floristic terms may be mixed as in the statement above that "fynbos" is a type of shrubland of the Cape kingdom.

Here, Cape Kingdom is a floristic designation, while fynbos and shrubland are structural terms.   The biomes or major vegetation regions of the world will be discussed in Chapter VIII, Macroecology.

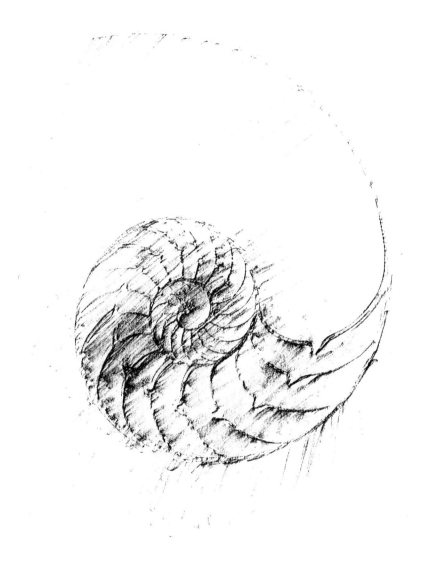

Figure 10. Cross-section of a nautilus (*Nautilus* sp.) (rubbing).

V

EVOLUTION

We see these beautiful co-adaptations most plainly in the woodpecker and missletoe [sic]; and only a little less plainly in the humble parasite that clings to the hairs of a quadruped or feathers of a bird; in the structure of the beetle which dives through the water; in the plumed seed which is wafted by the gentlest breeze; in short, we see beautiful adaptations everywhere and in every part of the organic world....

There is grandeur in this view of life with its several powers, having been originally breathed into a few forms or into one; and that, whilst this planet has gone cycling on according to the fixed law of gravity, from so simple a beginning endless forms most beautiful and most wonderful have been, and are being, evolved.

On the Origin of Species, Charles Darwin

Evolution

In 1831, a young Englishman, whose father wanted him to become
a clergyman, is asked to be the naturalist for the H. M. S. *Beagle* in
a voyage to South America to "survey the S. extremity," according
to its captain. His father is finally convinced the son should go,
after a letter from an uncle states: "The undertaking [traveling on
the *Beagle*] would be useless as regards his profession, but looking
upon him as a man of enlarged curiosity, it affords him such an
opportunity of seeing men and things as happens to few." On
October 2, 1836, Charles Darwin, a graduate of Cambridge
University (without honors), whose pursuits in 1828, a few years
earlier, had been listed as "collecting insects, hunting, shooting,
and being idle," returned to England after five years and two days
on the *Beagle*. These years dramatically changed a young man
whose observations and conclusions in time appeared in a work
entitled *"On the Origin of the Species by Means of Natural Selection, or
the Preservation of Favoured Races in the Struggle for Life."* Since its
publication in 1859, the way modern man looked at himself
and the natural world was forever changed.

One of the basic questions of science concerns chaos and order in
nature. Is order inherent in nature or does man order nature in
his perception of the world? Before Darwin, most philosophers,
theologians, and scientists saw intelligent design in nature. To
many, this design was the work of a higher power, the creative
force in nature. Since its publication, therefore, Darwin's work
has been considered by many (from the late 1800s to the present)
to be a challenge to religious explanations of man's creation,
especially the creation account in the Judaeo-Christian bible.
Some have even characterized Darwin as an anti-Christ bent on
destroying Christianity. Darwin, in fact, was fascinated and awed

by the creative force in nature. But, to him, this creative force was evolution by natural selection, not God. To Darwin, evolution was how nature got things done. It was, to him, the "economy" of nature to take something and make it better.

But Darwin's evolution is not quite the evolution of the "neo-Darwinists" of today. Darwin saw evolution as a long-term process, with only gradual changes to species taking place. Many modern evolutionary biologists have interpreted the fossil record in a manner that sees evolution proceeding rapidly at times and more gradual at other times. Darwin's gradualism has become "punctuated" equilibrium. Some present-day evolutionary biologists also point out that Darwin emphasized competition in nature, when cooperation may, in fact, be the driving force in evolution. Furthermore, theoretical scientists assert that life, being a complex system, contains inherent order, which over time changes naturally with or without disturbance or mutation, to other types of order.

Evolution—the noun is completely missing from *On the Origin of Species*—unlike taxonomy, biogeography, population biology, and even paleontology, is theoretical biology. Theories must be tested by observation and experiment. There are many interpretations of evolution –evolutionary biology probably creates more argument and discussion than any other subfield of biology. Most biologists, however, do accept the validity and universality of neo-Darwinian thought.

The study of evolution is the study not of mammals, sedges, or fish, but of how mammals, sedges, and fish change or have changed over time. Those who study evolutionary theory are historians of earthly biota. And as historians repeatedly "revise" our view of history, so evolutionary biologists reinterpret and reinvent the past. Evolutionary theory will, therefore, always be itself evolving.

Now add to the hundreds of books on evolution one 1433-paged giant of a book entitled *The Structure of Evolutionary Theory*, by Stephen Jay Gould. Published in 2002 shortly before Gould's death, the volume devotes about 500 pages to discussing Darwinism and over 500 pages to a "revised and expanded evolutionary theory." Gould's bibliography alone is 43 pages of 10-point type. With all due respect for Dr. Gould, who was one of my favorite modern essayist and was among the best science writers of his time, why does it take over 1000 pages to explain what "evolution" is? It is because evolution is more than just "science"; it is biological science's view of the history of life on earth, what it was when it began and what it is now. It is science, but it is also grand speculation by great thinkers; in summary, it is the current paradigm of biology. As Gould himself says, "No difference truly separates science and art.... We only perceive a difference because our disparate traditions lead us to focus on different scales...." *The Structure of Evolutionary Theory*, a book generally devoid of tables of data and illustrations, appears to be more of an historical novel than a work of science. It is a compendium/history of two men's views—Darwin's and Gould's, both scientists and both high priests of theory—of about everything there is and has ever been in the living realm of earth. And that is what the study of evolution is about.

Natural Selection

On winter mornings I often watch a Red-bellied Woodpecker (*Melanerpes carolinus*) forage for food on the bark of a pecan (*Carya illinoiensis*) tree in my backyard. The woodpecker climbs up and down the tree, methodically picking loose bark away discovering adult moths, bark beetles, spiders, roaches, and larvae of various species of insects. He meticulously works the tree from top to bottom at least once a day. He does not get all of the invertebrates on the tree, only enough to survive. There is a delicate balance between his foraging skill and the skill of camouflage and cryptic coloration invertebrates possess. If one maladaptive species of moth is more brightly colored than all of the others, and has no chemical or other defense, the woodpecker's ruthless foraging, along with that of other insectivores, would surely render it extinct.

How does nature take something and make it better? Darwin was fascinated by two subjects, as the first two chapters in *The Origin*—"Variation under Domestication" and "Variation under Nature," selection in plant and animal breeding done by man, and the variability of a species in the wild. Just as man could "select" for certain characteristics in plants and animals in agriculture, nature could do the same naturally. Darwin noticed (and discussed the fact in depth in "Variation under Nature") that individuals of a "species" in the wild were highly variable. This variability, of course, was a curse to systematists and taxonomists, who wanted species to fit neatly into their descriptions, but to Darwin, this natural variation was the raw material (or today, the "genetic material") for the process of what Darwin and A. R. Wallace called "natural selection." [Wallace independently arrived at an evolutionary theory about the same time as did

Darwin (they jointly presented a paper on the subject to the Royal Society).] According to Darwin , organisms change over time through "natural selection," a process that preserves "favourable" variations and rejects "injurious" variations. Evolution acts at the individual level. Changes that benefit the individual are adaptive. In individuals and populations of individuals where maladaptive characteristics (or genes) become more frequent, extinction gradually takes places. In short, adaptive offspring survive; maladaptive offspring do not.

Speciation and Extinction

Geographic separation of populations of a species usually drives the process of speciation. Geographic isolation results in little or no gene flow from the isolated population back into populations in the main range of the plant or animal, and *vice versa*. Isolating "barriers" are usually large—oceans, mountain ranges, deserts. The process of the birth and extinction of species that are geographically separated is usually a slow process taking thousands of years. Thus the red maple of China (*Acer faberi*) and that of North America (*Acer rubrum*) were probably once one species, but separation and evolution over thousands of years has isolated the two to the point where they no longer are capable of interbreeding and producing viable offspring.

It has been proposed that Edna's trillium (*Trillium persistens*), one of our subtitular species, was once populations (or a population) of the more widespread large-flowered trillium (*Trillium grandiflorum*). When the North American glaciers began to melt, about 20,000 years ago, and the Southeast began to warm up, large-flowered trillium and many other herbs of the eastern deciduous forest began to expand northerly with the deciduous forest to cooler environments. Edna's trillium may have been left behind on the rich soils of cool mountain gorges in northwestern South Carolina and northeastern Georgia. Isolation from its parental species reduced the gene flow from these populations into and out of other populations of large-flowered trillium, combined with random genetic drift (see below), may have resulted in Edna's trillium, a new and distinct narrow endemic species in the genus *Trillium*.

In environments where there are adjacent soil type differences (in plants) or microhabitat or niche differences (in animals), recent research, however, has shown that evolution may take place more rapidly. A species of grass growing on tailings from a copper mine quickly developed metal tolerance, while nearby individuals of the same species did not. Although growing adjacent to each other, these grasses have become so different in just a few generations as to warrant separate species status. Many other examples new plant species that have arisen on isolated soils types within the range of a parent species are now known; most of these are thought to have species much more rapidly than was historically thought possible. In animals and plants capable of long distance dispersal, however, what is termed the "founder" principle may be important in rapid speciation. In small, isolated populations, such as a small population of birds that have colonized a small island or a population of plants generated from one seed source, the total range of genetic differences is much less than that of the mainland or source population. Here, random changes in gene frequencies—a process called "genetic drift"— can be a powerful selective force. Because there is little mixing of genes from other populations of the species in question, new species may be rapidly formed through genetic drift.

Natural selection is a driving force in speciation. Over time, species gradually change due to minor differences in genetic composition that make some species more "fit" in a particular environment. The individuals with the fittest genes, therefore, survive—hence, "survival of the fittest." Over time, the resultant individuals may become incapable of interbreeding with the individuals of the source population. Genetic differences between the two populations become greater, and the fitter population may become a new species and the old population's species may become extinct. Or, the variant individuals from the source population may invade an uncolonized niche (such as the soil type islands mentioned above), become a new species, while the source population remains a viable species or becomes extinct due to competition, habitat change, or some other factor.

Adaptive Radiation

Nothing excites me as a naturalist more than to be told that there are over 600 species of the genus *Rhododendron* in the mountains of southwest China or that the Hawaii Islands have 300 species of the fruit fly genus *Drosophila*. It is not that I want photograph or collect each rhododendron or examine the eye color of each fruit fly species. What excites me is the idea that such diversity and species richness can and does exist in nature. And this diversity also validates my choice of being a naturalist and writing about natural history—it makes me realize that one can never get bored studying nature.

How evolution has created what Darwin called "endless forms" can best be seen in living groups of organisms that have radiated from a single or group of species into numerous diverse species. Darwin's finches of the Galapagos Islands all came from one species of finch that colonized these offshore islands from the continent of South America. Over time and isolation on separate islands, new species of finches, all with the same ancestor, arose. When these finches came back in contact with each other (when two or more finches occupied the same island), Darwin noticed that no interbreeding took place. Different courtship rituals had evolved in each species, and different niche utilization and feeding behaviors had also evolved. This process is called "adaptive radiation." An ancestral species produces new species by the process of niche utilization or niche specialization. In other words, individuals that use the resources of available local niches or microhabitats gradual become different from their congeners, like the finches, and ultimately become new, ecologically or geographically isolated species.

In the sedge genus *Carex* (Cyperaceae) in North America and Asia, hundreds of species have arisen (over 500 species in North America and undetermined number of species in China) by the invasion of microhabitats or niches by a few ancestral species. In a southern Appalachian gorge, for example, one species of *Carex* may grow on dry rock walls, one species may be found in seepage lower down in gorge, one grows in the spray of a waterfall, one is in the deep muck of a bog at the base of the falls, one in rich, limestone-derived soil on a slope near the falls, and so on. Each niche is like one of islands of the Galapagos; it is separated from the other niche by some feature that prevents interbreeding of the sedges. There are so many species of *Carex* in some areas that they can be used as indicators of minute differences in soil moisture and pH levels.

As Darwin recognized and we now understand more completely, adaptive radiation is the process that has dramatically increased the biodiversity of the earth. Some of the case histories are so impressive as to be difficult to comprehend. In northwestern China and in the eastern Himalaya, there are over 600 species of the woody shrub genus *Rhododendron* (Ericaceae). Here, the genus has produces hundreds of new species adapted to every possible available niche in this mountainous region; in each species a difference has been "selected for" that allows this species to do better in that niche than any other species. The marine mollusk genus *Conus* has around 500 species that have radiated around the world in different microhabitats. And the fly genus *Drosophila* has more than 300 species in the Hawaiian islands alone.

In Lake Malawi in eastern Africa, closely-related fishes of the family Cichlidae have evolved different head shapes, mouth functions, and feeding habitats (Figure 11). Some fish have mouthparts adapted to scraping algae off of rocks, some for plucking aquatic plants from the lake bottom. Some fish are carnivorous and have mouths that have evolved to eat zooplankton, some have mouths for crunching insects, and some prey on each other, being specially adapted to bite the others fin, scale, or eye. In most of these cases of adaptive radiation where many species have arisen from a common ancestor or group of ancestors, it is easy to see how habitats and microhabitats (niches) are used by each species. In many cases, however, the complex relationship between plant and soil or animal and plant or insect

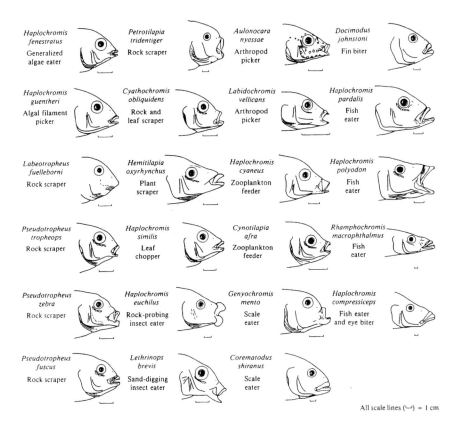

Figure 11. Evolution of mouthparts and feeding habits in cichlid fish in Lake Malawi (from Brown and Gibson, 1983).

and plant is not so easy to understand. "Let it be borne in mind how infinitely complex and close-fitting are the mutual relations of all organic beings to each other and to their physical conditions of life."

One of the most interesting products of the process of adaptive radiation is the phenomenon known as "convergent evolution." Unrelated species from distant locations have evolved similar growth forms, feeding behaviors, and appearances in response to similar ecological niches.   The sugar glider of Australia, a marsupial, fills the same ecological niche there as does the flying squirrel, a rodent, in North America.   The pythons of the Palaeotropics (Old World Tropics) and the boas of Neotropics (New World Tropics) are different taxonomically, but both groups are large constricting serpents living primarily in trees. In Africa where the succulent cacti (Cactaceae) are absent, succulence has evolved in the lobelia family (Lobeliaceae), non-succulent herbs in North America.   The orb web in spiders has evolved twice in separate families—Aranidae and Uloboridae.

There is the apocryphal (and probably true) story about a graduate student wrote a paper on the method by which two plant species had evolved and submitted the paper to the journal *Evolution*. Both of the journal's reviewers rejected the paper with the comments "pure speculation—needs hard data." The same paper, upon being submitted to the same journal, at a later date, by a famous evolutionary biologist, was pronounced "brilliantly insightful."

All of the above in this chapter having been said, it is true that much of evolutionary thought is still grand speculation. The hard data of evolution comes from the incomplete fossil record and scattered field observations on thousands of species. The study of evolution is not and can never be an exact science. At any one time on our planet, many individuals of many species are all competing with each other for food, territory, and other resources. Males are competing with males for females; females are competing with other females for males. The characters and behaviors in each species that prove to be most fit are somehow expressed in genes that are recombined through sexual reproduction, "the masterpiece of nature," according to Erasmus Darwin, Charles Darwin's grandfather. We, one species at one moment in time, in one glance, think that we can understand nature, which is itself always in flux? A thousand questions lead us to a million more. E. O. Wilson has said that our greatest biologists derive their greatness from asking the most interesting questions, not providing the most exact answers. Why does the raccoon have a penis bone and man does not? What is the function of the panda's thumb? Why do bamboos wait so long to flower? What is the meaning of virgin birth in nature? And so on.

Figure 12. Tasmanian scallop (rubbing).

VI

POPULATION BIOLOGY:

SIZE, LONGEVITY, AND LIFE HISTORY

Some seed the seasons mar,
Some the birds devour,
But here and there will flower
The solitary stars....

"I Hoed and Trenched and Weeded," A. E. Housman

"...that time allows/In all his tuneful turnings so few
and such morning songs/Before children green and
golden/Follow him out of grace."

"Fern Hill," Dylan Thomas

Population biology..."is concerned with the numbers of
organisms and with the consequences of those numbers
with birth and death rates, immigration and emigration,
with the consequences of exponential growth rates, the
process of colonization and the stresses that result from
overcrowding."

Population Biology, J. L. Harper

Introduction

Population biologists attempt to model and describe the temporal and spatial structure of populations and individual organisms. Population biology studies the demographics of plants and animals, the life histories of species, and the death of species. Some of the questions of population biology include: How does the size of the individual of a species affect its populations? Why do some organisms live short lives while others live long lives? How do populations use niches? How does competition and cooperation affect the life history of a species? Why are some species more social or more colonial (in plants) than other species? How does genetic differentiation and individual differentiation correlate with each other?

Size

The Great Barrier Reef, a two thousand mile-long reef of colonial coral animals, was for a long time thought to be the largest living thing on the face of the earth. A recent report from European biologists, however, sets the size of a massive colony of the introduced Argentine ants (*Linepithema humile*) at over 3,000 miles long, stretching from Italy to Spain. The blue whale (*Balaenoptera musculus*), at up to 26 meters long and 100 tons, is probably the largest non-colonial animal that has ever lived on the earth. The giant sequoia (*Sequoiadendron giganteum*) is the largest single-stemmed plant ever known. The giant albatross (*Diomedea exulans*) has a wing-span of around three meters, which makes it the largest living bird (or flying creature). [Note, however, that the now-extinct pterosaur Northrop's quetzalcoatlus (*Quetzalcoatlus northropi*) ranks as one of the largest winged creature ever to live (see Figure 5 for perspective on this creature).]

The sperm whale (*Physeter macrocephalus*) has the largest brain of any creature. Electric fish of the genus *Mormyrus*, however, have the largest brain to body size ratio of any known living creatures. And the giant squid (*Architeuthis dux*), at up to 20 meters in length, is the largest known invertebrate (it's a cephalopod, related to the octopus, the nautilus, and, more distantly, mollusks) and has the largest eyes—about 40 cm in diameter—of any living creature.

Like Jonathan Swift's old rhyme about little creatures that have littler creatures that bite 'em and *ad infinitem*, within the gut of most termites lives the smaller protozoans that to help the termite digest the cellulose in wood. Small creatures, however, like the termite (who can destroy houses), are not insignificant in the great scheme of things. Minute dinoflagellates only a few millimeters long (*Gonyalaux* sp.) that produce the deadly "red tide" are so numerous that they appear on imagery taken from the space [as have the microscopic shrimp-like krill (*Euphasia superba*) that occur in dense enough colonies to make portions of the subantarctic oceans pink].

Age/Class Structure

Population biologists are often concerned with the age/class structure of a population of animals or plants. The age/class structure may say something about the breeding system of the population or its ecology. In some populations of painted turtles (*Chrysemys picta*), for example, all of the individuals present are either very young or very old. Middle-aged individuals are absent. Here, predation is highest when turtles reach early middle age. If they can make it through this precarious period, they can survive to old age, when they have fewer predators because of their size.

Some plant populations have no seedlings, yet somehow seem to propagate themselves clonally until a major environmental disturbance occurs. When this happens, either massive seed production occurs (see the discussion of bamboo flowering below) or seedlings are recruited from an existing soil seed bank. Many species exist only as soil seed bank species until a major disturbance occurs, and they finally get their chance to germinate. Research has shown that the soil seed bank usually holds many more species than are present as above-ground plants. Green-and-gold (*Chrysogonum virginianum*), a spring-blooming composite (Asteraceae) found in eastern deciduous forests in North America, is rare or nonexistent in deeply shaded woodlands. But if you were to go out and chop down a tree or two and create a small light gap with a little soil disturbance, you would see the plant flowering in a few years. The seeds were dormant in the soil. Furthermore, its seeds are dispersed by ants, who will in a few years, move the plant around to all available niches in the light gap.

Longevity

Death is as much a part of the world in which we live as is life. Our bones one day will glow with phosphorescence in the dust whence we came. Carl Sagan and others have said that our very bone marrow is "star-stuff," elemental residue from the Big Bang.

And that eerie nocturnal green glow reported from old graveyards and woodlots where animals have died comes from the phosphorous in decaying mammalian bones was once part of the complex actions of the phosphorous in our DNA and RNA.

Some species live for years; others for only days. Humans and the giant tortoises of the Galapagos Islands live for over 100 years; elephants may live up to 70 years. The average life span of an adult butterfly is about seven days. Some adult mayflies live less than a day. Biologists are constantly debating over what is the oldest living thing on the earth (Table 3). Candidates include the gnarled bristlecone pines (*Pinus longaeva*) of the western United States; the mega-clones of aspen (*Populus tremuloides*), all stemming from one individual that germinated thousands of years ago; the creosote bush (*Larrea tridentata*) and its never-ending circular growth form; the clonal King's holly (*Lomatia tasmanica*), a shrub estimated to be 43000 years old; giant coral reefs such as the Belizean reefs and the Great Barrier Reef with millions of coral animals living as one unit; the giant sequoias (*Sequoidendron giganteum*) of the Sierra Nevada of California; and age-old clones of the honey mushroom (*Armillaria sp.*) recently found in Michigan and Oregon. A crustose lichen on an old boulder in a gorge in the southern Appalachians or on a cliff-face in the Sierra Nevada may be over 1000 years old; a bald cypress (*Taxodium distichum*) in a blackwater swamp in North Carolina has been dated as over 2200 years old; and some seasonal herbaceous perennials in temperate deciduous forest may have been around for centuries. The common bamboo (*Bambusa vulgaris*) lives to 117 or so years and then dies. Sometimes a clone of bamboo will cover several hectares before flowering, fruiting, and then dying.

Table 3. Long-lived biota.

Scientific Name	Common Name	Taxonomy	Growth Form	Max Age
Armillaria ostoyae	honey mushroom	fungus	clonal fungus	2200+
Geochelone nigra	Galapagean giant tortoise	reptile	turtle	100+
Homo sapiens	human being	mammal	hominid	125
Larrea tridentata	creosote bush	vasc. plant	clonal shrub	12000?
Lomatia tasmanica	King's holly	vasc. plant	clonal shrub	43000?
Populus tremuloides	aspen	vasc. plant	clonal tree	1000s
Pinus longaeva	bristlecone pine	vasc. plant	tree	4500
Rhizocarpon geographicum	lime-green map lichen	lichen	clonal crustose lichen	1000s
Sequoia gigantodendrron	giant sequoia	vasc. plant	monopodal tree	3200
Taxodium distichum	bald cypress	vasc. plant	monopodal tree	2200

Interestingly, when a clone dies, all of its shoots, be they in England, Japan, or Africa, will simultaneously die.

Individuals, Clones, and Societies

In population biology, before studies can be conducted of individuals, colonies (or clones in plants), or populations, the individual must be defined. In the trembling aspen or the coral reef mentioned above, defining the individual is often very difficult. Plant population biologists use the terms "genet" and "ramet." A genet is a genetically distinct organism, while a ramet may be an individual sprout or shoot off of a large genet. The distinction is important. In a woodland valley, all of the "plants" (ramets) of a species may have originated clonally from one seedling. If a pathogen spreads through the population, all of the ramets will be equally affected and may all die. It is good to know such things when studying natural populations.

In animals, the individual is usually very distinct, a "genet" in plant terms. In social insects and coral reefs, however, the entire colony may act as an individual. Corals may actually be clones of each other. And, in social insects, the colony is closely related, all of the workers being sisters. It is well-known that workers will sacrifice their lives for the queen and other workers, to whom they are more closely related than the males of the colony. (The males result from fertilized eggs, the workers come from unfertilized eggs.)

Competition, r, and K Selection

The idea of "everything is everywhere, the habitat selects" may be true for many spore-bearing, wind-dispersed microorganisms, orchids, and fungi, but for most plants and animals, habitats are limited and are often filled to capacity. A basic concept in biology is that two species cannot occupy the same niche. The plant or animal that possesses the best competitive strategy is the one that survives in a given microhabitat. To avoid being choked out of an environment, a plant or animal may express a phenology different from its competitors. Or it may colonize microhabitats such as rotting logs where other creatures invade only soil.

Historically, population ecologists viewed species competition in simplistic terms. Species that invaded habitats and grew without stress or regulation were said to be "r" selected species. (The letters "r" and "K" came from early population theory equations.) They include pioneer and early successional plant species and animals that invade unfilled niches. "K' selected species were said to be species that were prevented from rapid growth and expansion by competitive stress or by lack of available resources. These species were thought to be more stable, because they handle competition, and persist much longer than "r" species.   In a forested environment, an "r" species may be a pioneer plant species that invades a light gap on bare soil [such as the green-and-gold (*Chrysogonum virginianum*) mentioned above]. The "r" species should persist until its habitat (or resource, if it is an animal) is depleted. In other words, after the light gap closes shade-tolerant species will take over. In the same forest, a "K" selected species may be a sugar maple (*Acer saccharum*) or a beech (*Fagus grandifolia*), species that are competitive and shade tolerant and are going to be around for a few hundred years. Coast redwood (*Sequoia sempervirens*) and tulip poplar (*Liriodendron tulipifera*) (Magnoliaceae) were said to be both r- and K-selected. Coastal redwoods germinate in full sun in wet gravel beds and continue to grow and dominate as the forest ages; in eastern North America tulip poplars invade light gaps and old fields as pioneer species but persist for decades to become part of the canopy of old-growth forests. "Fugitive" species may even be shorter-lived that "r" species. These are plants or animals that are constantly on the move, colonizing niches that exist only for a short period in time. Fugitives include birds that colonize newly-created volcanic islands, and plants such as garden weeds that invade disturbed soils where there are no plant competitors. Over time population studies have shown that the "r" and "K" dichotomy was, like most dichotomies, an oversimplification of nature. In reality, there are groups of species that exhibit different growth "strategies" in nature, but these strategies are probably much more numerous and diverse than that which has been enumerated in the literature.

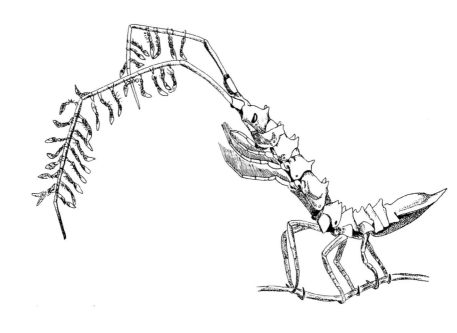

Figure 13. Female marine isopod (*Arcturus baffinii*) carrying
young on antennae (from Barnes, 1968).

VII

MICROECOLOGY:

TERRITORIALITY, SEXUALITY, AND

COEXISTENCE

All's Fair in Love and War

Microecology here refers to species interactions on a small scale; ecologists often use the term "organismic" ecology for what I am talking about here. (The next chapter will deal with macroecology or "systems" ecology.) This chapter could easily have been called "fighting, fornicating, and getting along." The fight for territory, the drive to reproduce sexually, and the energy that goes into peaceful coexistence are among the most important elements in the life of most organisms, including our own *Homo sapiens*. Territoriality can reveal the most base animal instincts in man and other species, but in its greatest moments, it can also be seen in love of homeland and love of country. Likewise, sex is a biological drive can span the gamut from simple reproduction to, in humans, to the love of another individual. And peaceful coexistence, which seems to be more common in nature than in human societies, can be everything from *laissez-faire* to mutualism.

Territoriality

"Good fences make good neighbors." Robert Frost

Man, dog, bird, and other beasts are territorial animals. Much of biology is fighting and warmongering for territory. In our species periods of peaceful coexistence are rare, as historians often point out. From an anthropocentric view of nature, nice guys do finish last. Those individuals (and species) who do not aggressively and successfully defend their territories get the worst territories, the worst mates, leftover food, and are, for the most part, no longer with us (see final chapter for a discussion on human territoriality).

Canines and felines are famous for "marking their territory." Even spayed female cats mark territories when other spayed females are around. Most of us, however, do not think of butterflies marking territories. But....I have a small cabin in a deciduous forest in the southern Blue Ridge mountains of eastern North America. Here, along my dirt driveway, a red admiral (*Vanessa atalanta*), a small, bright nymphalid butterfly, annually sets up a mating territory every summer. The red admirals prefer large light gaps in forests, and this road fits the bill for prime red admiral habitat. In these light gaps, the males advertise their colors with rapid wing beating to passing females. They daily patrol their territory, which is often less than 0.1 ha, and attack anything that enters. Red admirals, who have only a 5 cm wingspan, have been known to try to chase away birds, hawks, vultures, and even people from their territories. Sometimes when I walked through the red admiral territory, the male would dive by my ears, often tweaking one after several passes. The most amusing attacks, however, were after I bought a red pickup truck. There was something about the red truck that really irked the red admiral those summers. To the male, the truck was a giant red intruder, competing with him for female attention. When the truck entered his territory, he would dive violently at the hood and often bounce off. Fortunately for him the truck was only in his territory for a few seconds; I was never such a sadist as to park the truck there to see what he would do (he might have done nothing because it was not moving or he may have mutilated himself as in the following story).

Similarly, during mating season, the male cardinal (*Cardinalis cardinalis*), a bright red passerine bird found in eastern North America, has been known to repeatedly attack their image in mirrors and reflective windows until the window or mirror is completely covered with red bird feathers. Large jumping spiders (salticids) often defend small territories with direct eye contact (they have the largest eyes of any spider) and the waving of their forelegs. I received a call from a friend once who said that there was a large hairy "thing" on her porch, staring her down in an aggressive pose. She was afraid and wanted me to come and move it so that she could exit her porch. It turned out to be a 3 cm long *Metaphidippus audax*, the so-called "bold" jumping spider. With aggressive territorial behavior, embodied in waving movements of its forelegs, a creature just over one inch long had stared down a mammal 180 cm tall (about 6 feet).

Interestingly, in nature, killing is often not cost-effective in terms of ecological energetics. And most territorial fighting in species other than man and a few other weird primates does not end in death. It simply takes too much energy to kill the competitor. In stag beetles, the giant mandibles are used to toss the "other" male as far away as possible so the conquering male will have no competition (usually in sex). In elephant seals, the battles are tougher, but no one gets killed. The same is true of birds and my red admiral butterfly discussed above, who probably couldn't kill anything, even if he wanted to. Mammals often exhibit a behavior called "submission" that saves a good bit of time and fighting. The lesser male, who wants the dominant male to know that he poses no threat, exposes his neck to the dominant male or, frequently, (as did my neighbor's small Jack Russell to my older, larger beagle) the lesser male falls on the ground with legs up.

This message is quickly understood by the dominant male who realizes the territorial intruder is no threat.

Territoriality is often closely associated with food source or food supply. If your preferred food is the nut of the jubjub tree, you must hang out near the jubjub tree, and you must guard the jubjub tree. But much of territoriality also derives from guarding territories in which a desired female occupies or will occupy. In other words, sex and territoriality are often reciprocally important in species interactions. "Sex" in nature is much more than just an act. It is color and morphological difference, courtship, coitus or copulation, and parental care.

Sexual Coloration/Dimorphism

In the human world, we usually think of the female as the most beautiful sex. In truth, however, in most species, the male is the fairer. In birds, the brighter male finds a territory and then begins displaying his colors to the female. In warblers, for example, the males are brilliantly colored, while the females are drab and nondescript. In fish, the male also rules in beauty. The cheek of the redbreast sunfish (*Lepomis auritus*) is vermilion; that of the female is a dull yellow. The males of many butterflies, such as the swallowtails (Papilionidae), are brightly colored while the females are black or dark-colored.

Mimicry is important in the coloration of the females of many species of butterflies. In eastern North America, the female tiger swallowtail (*Papilio glaucus*) is black, mimicking the distasteful pipevine swallowtail (*Battus philenor*). In the Diana (*Speyeria diana*) (Nymphalidae), the male is a bright orange, while the female is black, again mimicking the pipevine swallowtail. Interestingly, in the Atlantic Coastal Plain where the pipevine swallowtail is rare or nonexistent, the tiger swallowtail female is yellow, matching the male's coloration. Mimicry is much more widespread in the tropics, especially in the Neotropics, where in the mid-1800s, H. W. Bates' observations of mimicry in Amazonian butterflies are among the most fascinating and illuminating field studies ever conducted in biology.

In larger animals, males are usually about the same size as the females and are rarely larger than the females. In invertebrates and some fish, however, males are usually smaller or much smaller than the female. It has been observed that in deep-sea marine fish, the males often got proportionately smaller with depth. The cephalopod (*Argonauta hians*), often called the "paper nautilus" because of the resemblance of the female's egg case to the true nautilus has males that are as small as 1/18th the size of the females. Male-female size dimorphism is also strikingly noteworthy in spiders. The males of the genus *Nephila*, a genus of large, tropical orb-weavers, are frequently 1/10th the size of the female and, rarely, as small as 1/20th the female's size. This size difference makes mating, under the best conditions, very difficult. The male, after competing with 10 or so other males in the female's web, must slide down the female's drying thread (the thread she hangs on after her last molt) to safely mate with her. In the tropical orb-weaver genus *Micrathena*, the male is also much reduced, is secretive, and often lives only a few weeks (see Figure 14). The male of black widows (*Latrodectus* spp.) is usually about 1/3rd to 1/4th the size of the female. Anecdotal information has the male always being eaten after sex; the truth is he is often eaten

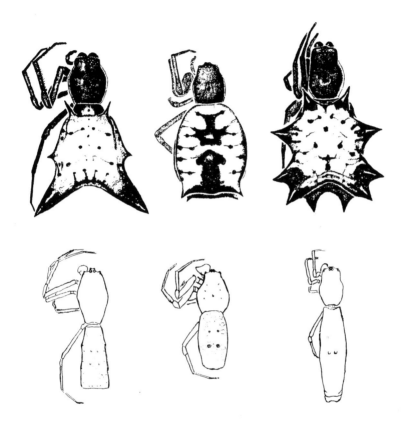

Figure 14. Females (top row) and males of the spider genus
Micrathena. Left to right: *M. sagittata, M. mitrata, M. gracilis*
(from Gaddy and Morse, 1985).

before sex—there are always plenty of other males. To the female, a male may be just prey in the web.

Sexual dimorphism is most notable in social insects, especially the ants. Here, in some species, there are castes of all-female workers with major, media, and minor workers (based on size). Reproductive individuals are drones (males) and queens, which are the winged members of the colony. In some ants, a queen with eggs may be over 50 times larger than a minor worker.

Courtship

In passerine birds, the male selects the territory and waits for the arrival of the female. The female selects the male with the most "attractive" territory. In bower birds in New Guinea, males build elaborate, domed nests and lay out colorful articles to attract females. *Anolis* lizards use brightly colored throatfans or dewlaps to attract females of their species (and to ward off other males from their territories). Male bullfrogs (*Rana catesbiana*), like males of other frogs, birds, and insects, use song to attract females. It has been shown that the bullfrog with the loudest song usually attracts the most females. Jumping spiders attract females by waving their forelegs in a complex pattern, mesmerizing the female. Loons and other birds are famous for fascinatingly elaborate mating dances. Male silk months (Saturnidae) fly for miles in search of the female source of the pheromone they have detected with their intricately-designed antennae. In lampyrid beetles (fireflies), males signal females of their species, some of which signal back during courtship. One predatorial species has short-circuited the system by learning to signal to males of another species who approach the larger females and become her prey instead of her mate. It has been reported that trout quiver violently with their mouths open while engaged in sex, followed by the supposed simultaneous release of egg and sperm. Swedish scientists found that males always released their sperm at the end but that, half of the time, the females cheated, withholding their eggs. The researchers speculated that the female was "faking an orgasm" as a mating strategy to preserve the eggs in case she ran into a more desirable male. Courtship in social insects involves a spectacular "nuptial" flight during which the males and females first meet in mid-air. In the spring, small piles of dead male carpenter ants (*Camponotus* sp.), the leftovers of the nuptial flight, are often found in leaf litter near hollow trees.

Sex, Bizzare and Otherwise

Getting back to the male of the tropical spider *Nephila*, who we left hanging on the female's drying thread, we must figure out how the minute male is going to inseminate the giant female. In spiders, this is done by external fertilization. The spider has two bulbous like structures on modified legs; he manually fills these pressurized containers with sperm. When he approaches the female, he empties the sperm reserve with syringe-like action into her epigynum (vagina/ovary). Different species of spiders have different insemination angles and methods, as attested by Figure 15, a "Kama Sutra" for spiders.

Figure 15.   Sperm insemination approaches by spider males
(from Emerton, 1901).

In bedbugs, the male injects sperm directly into the female's bloodstream, a technique called "traumatic insemination." In dragonflies, mating usually takes place on the wing. The raccoon, your family dog, minks, bears, and some marine mammals have a special bone called the baculum in the penis, which prevents the loss of an erection a critical moment in the life of the animal. Sexual dynamics in gastropods (Phylum Mollusca) is quite bizarre, compared to that of other creatures. Many gastropods are hermaphroditic, harboring both male and female parts. This sets the scene for mutual copulation in which a snail may insert the penis into the vagina of another snail while the other snail is inserting its penis in the first snail's vagina. Such chains of copulation are often formed in some gastropod species. In the marine slipper shell genus *Crepidula*, young individuals are always males. As a male ages, he will remain a male as long as it is attached to a female. Once it becomes independent from the female, it, too, will become a female and will remain so for the rest of its life. In the plant Jack-in-the-pulpit (*Arisaema triphyllum*) (Araceae), the single flower produced annually may change from male to female from year to year. After years of heavy fruit production, the flower produced the following year is usually male.

Most creatures are careful about where they put their sperm, but in mollusks, algae, most corals, some vascular plants and crustaceans, sperm (pollen) is set adrift by males, only to meet with eggs at a later moment. In coral reefs, coral polyps often release sperm in unison a few nights after a full moon, clouding the seawater for several hours.

Finally, some creatures have no sex at all. It is thought that around 1000 of the 1.5 million or so species on the face of the earth are female clones that have reproduced without the benefit of males. This form of reproduction, called parthenogenesis, has been found in aphids, lizards, and snakes.

The sexual dynamics of plants may seem simpler than that of animals; however, sex in plants in extremely complex. Most flowering plants have perfect flowers, flowers with both male (staminate) and female (pistillate) parts. To prevent self-fertilization, plants have evolved various methods of outcrossing. In some plants, the male parts become sexually active first, a system call protandry; while in other species, the females parts are receptive before the staminate structure release pollen, a condition referred to as protogyny. Some plants have only female or male flowers (dioecius), while some species have male flowers and females flowers on the same plant (monoecius). In the aster family (Asteraceae/Compositae), there are ray (outer) flowers and disc (central) flowers. In some species, such as the Russian mammoth sunflower, all flowers are fertile, but in other species (e.g., *Chrysogonum virginianum*), only the ray flowers produce fruit, the other flowers are just for pollen production. Dandelions reproduce by apomixis or bud pollination, a botanical form of parthenogenesis.

Pollination and the methods by which flowering plants attract pollinators are even more fascinating. Orchids have evolved flowers that look like bees, flowers that trick bees into simulated sexual behavior. Many plants have colors that attract hummingbirds. Other plants have deep nectar that can only be reached by certain long-tongued moths or hummingbirds. Some plants flower at night and are pollinated only by night-flying moths.

Parental Care

As anyone who has taken care of a baby for a year or so knows, parental care can be difficult and time-consuming. In mammals, birds, and many other creatures, female nurturing is usually the rule in nature. The female protects and feeds the young until they are ready to depart the nesting area. Female nurturing is even common, though it different ways, in invertebrates. The female green lynx spider (*Peucetia viridans*) and other spiders construct and egg case and remain on the eggs until they hatch or until the female dies. Young scorpions and wolf spiders live on the female's back (young scorpions are usually transparent) until they are large enough to forage on their own. Females of the marine isopod *Arcturus baffini* camouflage the young and carry them on their antennae (Figure 13). In ants, specialized workers or "nurses" carry the larvae from nest to nest and feed them in the nest. It is thought they also communicate to foraging workers what type of food is needed for the developing ants. In the green salamander (*Aneides aeneus*), the female lays her eggs on the roof of a crevice, curls up around the 10 or so eggs, and stays there until they hatch several weeks later.

Some sac spiders imprison the female in a silk sac for a few months after sealing her epigynum to avoid other sperm being placed there. Speaking of prison, the male of the paleotropical bird the hornbill (*Buceros* spp.) (Bucerotidae), mates with a chosen female, seals her in a hollow tree cavity with cellulose tissue, and leaves her to hatch and rear the young in captivity. The male must drill through the tissue every time he returns to the nest with food. He reseals the nest and then leaves to forage again. He first feeds the female, but later must collect additional food for the young.

Male nurturing is well known in the animal kingdom. Many fish species make the nest and guard the eggs and young while the female forages for food. In the giant water bug genus *Belostoma* the female cements the eggs to the back of the male where he carries them for about a week or two until incubation.

Coexistence, Competition, and Symbiosis

Ecologists assert that no two species may exist for an extended period of time and occupy the same "niche." "Niche" in biology is a concept that includes a species habitat, food requirements, and spatial relationships with other species. Theoretical ecologists speak of as "equilibrium coexistence," therefore, as the natural result of competition, which results in stability. But is equilibrium coexistence more an abstraction of mathematical ecologist than a reality that can be observed in the field. If you go into an old-growth forest that has never been logged, do you see giant trees growing, park-like, in lines? No, you find light gaps with dead tree boles, weeds, and young trees interspersed between areas of large trees. Nature is not so neat. Something is always going on. And no one comes behind to straighten up after the mess. Some have likened the orderly chaos of natural forest systems to a "shifting-mosaic, steady-state." The old-growth forest is a steady state system, but if you step into the wrong spot, it might not look like it from your point-of-view. Other natural systems are probably very similar to forest systems, but no one has yet to come along to explain them properly.

Mutualism, Commensalism, and Parasitism

A mutualism occurs when two species interact in a way in which both species are benefited. For example, tropical fish often have smaller fish that clean their body in return for the leavings from meals. Most hard corals have algae-like zooxanthellae that live in the coral polyp itself. (The zooxanthellae are actually the tailless non-swimming stage of a dinoflagellate, the same phyto-marine organisms that produce the famed "red tide.") The zooxanthellae benefit by having a place to live, while the coral gets nutrients from the algae. Ants are often involved in mutualistic relationships with tree species and other plant species. Many tropical species, especially in the family Melatostomaceae, have special ant-bodies for housing ant colonies that will ultimately protect the plant from herbivores. Myrmecochory, the disperal of seeds by ants, is a highly evolved form of seed dissemination that is beneficial to ants and to the plants whose seeds they disperse (Figure 16). It was originally thought that ant dispersal of seeds was rare in nature, but in the late 20th century, it was shown that myrmecochory is widespread in some community types. In the fynbos shrublands of the Cape Floristic Province in South Africa, for example, as many as 2500 species may be ant dispersed.

Commensalisms are defined as a relationship in which one creature benefits while the other receives no benefit or detriment. Some commensalisms may be undiscovered mutualisms (or parasitism). The very unglamorous branchiobdellid worms (Phylum Annelida) live on the gills of crayfish. They eat diatoms,

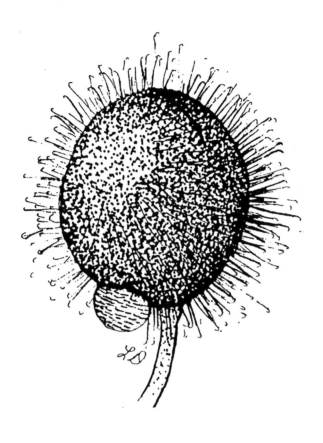

Figure 16. Elaiosome on ant-dispersed seed of *Galium circaezans* (Rubiaceae) (Lesa Dill).

algae, and organic debris. Considered commensals, could these creatures be cleaning the gills of crayfish? If removed from the crayfish, they live for months, but removed, they never lay eggs. Some propose that this group, which is closely related to leeches, sometimes suck blood. Maybe, therefore, as our knowledge of this interaction increases, we will conclude that the worms are actually crayfish parasites. And possibly even more fascinating is the interaction of two species of European caddisflies. Caddisfly pupae are worm-like aquatic creatures that build a pupal case of sticks, stones, and small debris in the bottom of freshwater streams. Only the head protrudes from the case, which is carried along the stream bottom by the pupa as it forages for vegetable matter. In the species *Limnephilus rhombicus*, the pupa's case is built not from sticks or stones but from smaller, living cases of the pupae of another species of caddisfly, *Beraeodes minutus* (see Figure 17).

It has been reported that the average number of species with which any species normally interacts is between 3 and 5. It has further been asserted that 5 to 10 parasitoids may be associated with each species of herbaceous-feeding insect. If this is true, much of nature is highly specialized parasitism. Parasitism is not just liver flukes; some of the most interesting insects are parasites. Parasitoid or solitary wasps parasitize many species of insects. Mud dauber species select one species of spider on which to live and they parasitize only that species. Mantis-flies (Mantispidae) parasitize spider egg cases; one species lays its eggs only on black widow (*Latrodectus* spp.) egg sacs. In pine savannahs in the southeastern United States, there is a wasp that collects one species of grasshopper, paralyzes them and neatly packs them

Figure 17. Larval case of the caddisfly *Limnephilus rhombicus* composed of the smaller cases of *Beraeodes minutus*, another species of caddisfly (from Majecki, 1983) (courtesy of John Morse).

with grass and eggs into the tube of the hooded pitcher plant, *Sarracenia minor.*

Detailed studies of the interconnecting matrix of life have shown that things are always more connected (and stranger) than we realize. DDT that was sprayed on the roofs of thatched huts in Asia killed its target—mosquitoes—*and* wasps that controlled the population of bedbugs. The mosquitoes were gone, but so were the wasps. In the end, the DDT created a bedbug problem. That big fish eat little fish and little fish eat smaller fish and *ad infinitum* has proved to be so true in the field of ecology. The butterfly that lands on the temple steeple does change the balance of nature; the one drop of rain falling into the sea changes sea level.

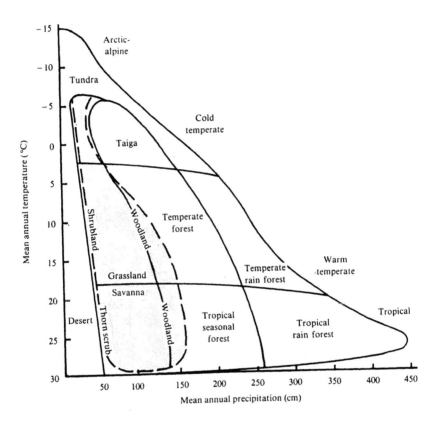

Figure 18. The relationship between climate and vegetation (from Whittaker, 1975).

VIII
MACROECOLOGY:
THE BIG PICTURE

Abiotic Factors: Climate, Landforms, and Soils

I once took a team of biologists to the mountains of south central Idaho. Most of us were from the eastern United States and were somewhat insecure in our knowledge of the western arboreal flora. We were surprised to find out that there were only ten or so common trees in the portion of Idaho in which we were working (a small woodlot in the southeastern United States may have 15 species of trees). Why might there be fewer species of trees in central Idaho than in, say, North Carolina. Central Idaho gets about 30 to 50 cm (12 to 20 inches) of precipitation per year (mostly snow); the mountains of North Carolina get 112 to 250 cm (45 to 100 inches) each year. The obvious answer is water—that substance essential to life; there is less moisture stress in North Carolina, and trees need moisture.

Which abiotic forces, therefore, exert the strongest influence on the variety of life from place to place. The single most important force is climate. Life cannot exist without water. Without precipitation or with very little water, plant and animal growth is severely limited, as in deserts. Another delimiting feature of climate is temperature. Most plants cannot grow in frozen polar regions. [Most of the animals that survive in these regions (penguins, marine mammals, polar bears) get their food from the oceans.] The areas on the globe with the most luxuriant vegetative cover are, therefore, warm, wet areas—the tropical rainforests.

Landforms and soils are also important delimiting abiotic features. Just as plants cannot exist in polar lands, plants and plant-eating creatures are severely stressed in mountain regions. Because temperature decreases with elevation, the world's highest mountains are barren ice fields analogous to the polar regions. Soils or the lack thereof can also limit the distribution of plant and animal species. Rocky areas support different communities of plant and animals species than areas with deep, rich soils. Where rainfall is adequate, mineral-rich soils, especially those rich in calcium and magnesium, often produce rich forest communities. Some soil types, however, limit species richness. For example, few plants will grow on serpentine- or nickel-rich soils. Saline soils are well known for their specialized plant communities. It was originally thought that the plants that dominate in saline soils could not grow in nonsaline soils. Research has shown, however, that some saline-tolerant species *can* grow in nonsaline soils, but they compete well only in saline soils.

Vegetation Regions: Biomes

The term "vegetation" refers to plants as living forms or structures in the landscape; the "flora" of a region implies the taxonomic structure of a region's vegetation. I once asked a student in an elementary physical geography course what kinds of plants were found in grasslands. The answer I was looking for, of course, was "grasses," a taxonomic term. Understanding the concept of vegetation structure all too well, she replied, "little ones." The vegetational structure of the earth provides a framework within which other organisms exist. Although there have been maps of the zoogeographic regions of the earth, there are many more attempts at mapping the earth's plant geography. Most maps of the earth's vegetation are based on vegetational structure—tree, shrub, grassland, etc.—instead of the taxonomy of plants.

Vegetation regions, sometimes called biomes, are broad entities encompassing different forest types or different shrub types of one classification such as "forest" or "shrub." Generally, climatic variables are used to further define vegetation regions. We, therefore, have the terms "tropical rainforest," "temperate rainforest," and "mediterranean shrub." Below, eleven major biomes of the earth are discussed. Figure gives their global distribution.

Tropical Rainforest—In tropical rainforests, precipitation is usually greater than 250 cm per year. The land in this biome type is usually lowland plains or floodplains. Trees are broadleaved evergreen species and reach nearly 100 meters in height. Lianas, woody vines that grow from forest floor to the tops of trees, are found in every tree. Because the soils are wet and soft in lowland tropical rainforest, trees have developed elaborate buttresses to support their large biomass. Windthrow is still common and is the most common way the forest regenerates itself. Because the lower levels of the forest are very dark, a well-developed ground layer is not present in tropical rainforests. Air-plants or epiphytes are common on the trunks, limbs, and branches of other plants.

Tropical (wet/dry) Deciduous Forest—In many parts of the tropics, there is a wet season that usually coincides with the sun's highest angle (June in the northern hemisphere and January in the southern hemisphere) and a dry season occurring about the time of the sun's lowest angle. These regions do not have enough moisture to support lowland tropical rainforest. They are, therefore, dominated by tropical deciduous forests or, if the drought period is long enough, tropical deciduous savannas, which are discussed below.

Tropical Savannahs—The grasslands of the tropics, tropical savannas are generally delimited by the severity of the dry-season drought. Tropical savannas are dominated by tree species similar to those found in tropical wet-dry deciduous forests, where deciduous legumes are quite common. Fire reinforces drought in the maintenance of grassland in most tropical savannas. In African savannas, tropical savannas are perpetuated by a combination of drought, fire, and the destructive action of large populations of elephants.

Temperate Rainforest—In the temperate zones of the Pacific Northwest in North America, Tasmania, Chile, and New Zealand, needle-leaved evergreen forests (in the northern hemisphere) and broad-leaved evergreen forests (in the southern hemisphere) occur in cool, west coast marine climates where steady year-round rainfall and/or fog prevail.

Temperate Deciduous Forest—The temperate deciduous forest once nearly wrapped around the entire northern hemisphere stretching from eastern China to central Asia. Climatic changes in portions of the northern hemisphere have now isolated this type in eastern China, eastern North America, Europe, and the Caucasus Mountains. This forest type is seasonally green, dominated by broadleaved deciduous tree species. It is rare or nonexistent in most of the southern hemisphere.

Mediterranean Shrub/Open Savannah—This type is found in southern and middle California, the Mediterranean, central Chile, western Australia, and the Cape area of South Africa. It is generally dominated by broadleaved evergreen trees with open, grassy savannas and by shrublands called "garigue" (in Europe), "chaparral" (in California), or "fynbos" (in South Afrrica), according to where you are. Summers are dry and are subject to fire, while winters are wet.

Boreal Forest (Taiga)—The boreal forest is one of the largest expanses of forest on the face of the earth. Needle-leaved evergreen trees of the genera *Abies* and *Picea* dominate this forest, which is found primarily in North America and Asia. It is absent from the southern hemisphere.

Temperate Grasslands—Grasslands are usually found in the central portion of continents away from coastal waters. The vast steppes of Asia, the prairies of North America, and the pampas of Argentina are all famous grasslands. Grassland usually occur where rainfall cannot support arboreal vegetation (less than 50 cm of precipitation per year).

Alpine Vegetation—Alpine or high elevation vegetation is generally found above the tree line over 3000 meters. Woody vegetation found in alpine areas is stunted or flag-form due to cold winds and icy conditions. Some shrubs take on a creeping habit and invade shallow depressions and crevices. Herbaceous plants and grasses from various genera and families appear in tufted cushion form, a life-form that allows them to protect the terminal growth bud of the plant from dessication and freeze.

Tundra—Tundra plants are similar to those found in alpine areas; however, tundra plants must deal with permafrost, a permanently frozen state the soil lies in for much of the year. Lichens and mosses are more abundant in tundra areas than in other biomes. Woody plants are rare or absent in tundra. In summer extensive sedge wetlands support large populations of insects, especially mosquitoes.

Desert—Deserts vary in absolute dryness, but the general condition one finds in most of the earth's deserts is that of nearly year round drought, lack of woody plants, and, in some cases, lack of plants at all. Some deserts exist due to oceanic influences and have fog belts with scattered vegetation (Atacama Desert of Peru, Namibian Desert)); some deserts have brief periods of rainfall that support bursts of wildflowers and herbaceous growth (Sonoran Desert); and some inland deserts are so dry that they support little or no vegetation.

Of course, these types could be even more grossly generalized into six types: forest, savanna, grassland, shrubland, and desert systems. Or, to go in the other direction, they could be broken down into smaller plant *communities* and plant *associations*. [In the United States, The Nature Conservancy and the U. S. Forest Service have recently compiled a classification of the natural communities of the United States containing over 5000 association types.]

Biogeochemical Cycles and Global Warming

When one looks at the "the big picture" of life on our planet, a complex and dynamic earth is seen. The earth, of course, is powered by the sun, but just how solar energy interacts and drives climate, the oceans, the carbon dioxide/oxygen balance, and the major elemental cycles (carbon, nitrogen, sulfur, and phosphorus) is poorly understood. A few decades ago, proponents of Gaia, an organic mother earth theory, were the laughing stock of atmospheric scientists; it appears now that the living earth is much more in tune with the geophysical environment than traditional science previously thought. Today, some of their conclusions of those formerly considered on the edge of science, especially in the field of environmental feedbacks, have been taken much more seriously.

There is currently very convincing evidence—from long-term atmospheric data at several sites and from data from ice cores that record carbon dioxide levels for the past 220,000 years—that the amount of carbon dioxide in the atmosphere has dramatically increased, especially in the northern hemisphere, since the mid-1800s. If this increase continues, will the atmosphere trap heat and become warmer or will it become denser and precipitate global cooling. Because climate history seems to indicate that there are major environmental changes on earth every 20-40,000 years and every 100,000 years there is major cooling (ice ages) (and throw in an asteroid striking the earth every few hundred million years), our future is not clear at all (as if it ever were). Some have pointed out that we may be entering a short glacial period about the same time as carbon dioxide-driven global warming is entering the picture, two global incidents that may actually cancel out each other.

Figure 19. American ginseng (*Panax quinquefolius*) (Hu Ye).

IX

CONSERVATION BIOLOGY:

ABUNDANCE CHALLENGED

Tyger, tyger, burning bright
In the forests of the night
What immortal hand or eye
Doth frame thy fearful symmetry.

"Tyger, Tyger" by William Blake

Extinction

The fossil record is filled with evidence of the gradual and the sudden extinction of species, floras, and faunas (see Chapter IV), but what concerns us in this chapter are species that have been extirpated or almost extirpated by man. Here, we are talking about the Dodo, the Passenger Pigeon, most of the giant tortoises of the islands of the Indian Ocean, the honeycreepers of the Hawaiian Islands, the Great Auk, the Tasmanian tiger, the Eskimo Curlew, and many other species. In just North America alone, it has been estimated that since the coming of European man nearly 500 species of plants and animals have been extirpated by man and his activities In two very different study areas, the Hawaiian Islands and the Tennessee River drainage, one can observe the dramatic perturbations caused by the advent of industrial man.

The Hawaiian Islands were colonized by Europeans in the late 1700s. With the advent of white man came diseases, agriculture, rodents, rabbits, and mosquitoes. The native Polynesians, who had lived on the island for centuries, were fishermen who cultivated some taro (*Colocasia esculenta*) and grew coconut palms (*Cocos nucifera*) on a small scale. The colonists' were interested in plantation agriculture, which needed expanses of cleared land. European man, therefore, set out to make Hawaii into a chain of plantation of islands. This resulted in ecological disaster: over 250 native Hawaiian species, including over 70 species of birds and 50 species of terrestrial snails, are *known* to have become extinct within the last 200 years.

On another front, in the early part of the 20th century, the United States was looking for areas where cheap, hydroelectric power could be generated and transmitted to the large population centers of the east. The wild, sparsely-populated Tennessee River valley was the perfect place for such a venture. Here were deep valleys, gorges, and canyons where the river could be boxed in and controlled. In the early 1900s, the Tennessee Valley Authority, an agency of the federal government, was formed. By the mid-20th century, nearly all of the Tennessee River, from North Carolina, northeastern Tennessee, northern Georgia, north Alabama, and into Kentucky had been dammed. It was unfortunate that the Tennessee River was selected. As it turned out, Tennessee and Alabama have the richest freshwater fish and mollusk fauna in North America. Primarily because of the creation of the Tennessee Valley Authority, the State of Alabama is second to the State of Hawaii in the number of species that have been extirpated in the last 100 years: 73 Alabama species are now extinct, most of them fish and mollusks, 17 in the pebblesnail genus *Somatogyrus*.

Abundance and Rarity

In the study of conservation biology, one must first define rarity before that concept is discussed. Some species are naturally more abundant or rarer than others. Species' occurrence can be divided into four general categories:

I. Species that are widespread and occupy a nonrestricted habitat;
II. Species that are widespread and occupy a restricted habitat;
III. Species that are restricted and occupy a nonrestricted habitat; and
IV. Species that are restricted and occupy a restricted habitat.

A type I species is the dandelion (*Taraxacum officinale*). It is found all over the world and grows in many different habitats. It could be called cosmopolitan or globally abundant. An example of a type II species is the tussock or bristly-stalked sedge (*Carex leptalea*). This fine-leaved plant is found only in acidic bogs; however, it is found in this type of bog from Alaska to Florida. Its range is widespread, but its habitat preference prevents it from growing everywhere. Other type II species would be circumboreal plant species that are found only on rocky barrens or alpine, but have a broad range. This type of species is not common but neither is it rare. Its abundance depends entirely availability of its preferred habitat in a given region. A type III species would be the Christmas Island crab (*Geocarcoidea natalis*), which is found only on Christmas Island south of Java, but is everywhere on the island. Many endemics—especially insular species—may fall into type III. They are found only within a narrow geographic range, but within that range, they are quite common, or sometimes called "locally common." Type IV species are usually the rarest species. Their range and their habitat are restricted. A good example of a type IV species is the pool sprite or snorkelwort (*Amphianthus pusillus*) (Schrophulariacae). This small aquatic plant is found only in vernal pools on granitic flatrocks in the Piedmont of the South Carolina and Georgia. It occurs only in two states and only in a very rare microhabitat.

Declaration of rarity in species may often be subjective. A limestone-loving fern may be called "rare" in one state or region because limestone is generally absent from that region. On the other hand, the same species may be abundant in a nearby state that harbors abundant limestone habitats. A cave-dwelling beetle may occur in numerous regions, but it may never be abundant in any region because of its limited habitat. Obviously, the terms "rare" and "abundant" can be comparative, qualitative terms.

Extirpation of Species by Man

Extinction by Exploitation

The Dodo (*Raphus cucullatus*), a large flightless dove, has the distinction of being one of the most famous creatures that has been extirpated by man. The bird is dubiously celebrated in our term "dodo," which is used as an insult in our modern vocabulary. Dodoes were found on Mauritius, an oceanic island in the western Indian Ocean. They had lived there for thousands of years, in relative peace. Passing sailors often ravaged their colonies and killing the birds for food, but it was not until 1644, when the Dutch established a colony on the island, that the Dodo was threatened. Clubbing of the flightless, fearless Dodo seemed to be sport during the early days of the settlement. Add this to the introduction of dogs, cats, monkeys, pigs, and rats, all of which overpopulated the island with young of their own to such a degree that most human settlers even left the island. By the year 1680, the Dodo had been exterminated.

Sadly, the Great Auk (*Alca impennis*), a north Atlantic colonial penguin-like bird, suffered the same fate as the Dodo. Like the nesting habitat of the Dodo, that of the Great Auk was on major sea lanes. In the late 1700s and early 1800s, ships heading toward North America filled their larders with meat and eggs from the Auk colonies. Later, the Auk nesting islands were relentlessly attacked by collectors who herded the flightless birds into pens and threw their bodies into boiling pots. The feathers, which were used a substitute for eider-down, were separated from the boiling carcasses, and later sold. The last Great Auks were killed off the Icelandic coast in the mid-1800s. At one time, this magnificent *"pengouin"* (Welsh for "white head") had been found from the northeastern United States to Canada, Greenland, Iceland, and even in the Outer Hebrides of Great Britain.

Another dove of sorts, the Passenger Pigeon (ornithologists capitalize common names, botanists don't) (*Ectopistes migratorius*), or the Migrating Dove, was at one time one of the most numerous birds in North America. Some think that in the 1800s, forty percent of *all* the birds of North America were Passenger Pigeons. They traveled in flocks of billions. J. J. Audubon and Alexander Wilson visited separate flocks, each estimated at over one billion individuals. Another flock was reported to be one mile wide and 320 miles long. How could a species with so many individuals be exterminated? Like the Dodo, the Passenger Pigeon fed and nested in groups; therefore, if one could find one roost or flock, one could "collect" lots of birds. The pigeons were sold for their meat, down, and feathers. Truckloads and trainloads were killed by every method imaginable. In an 1878 hunt in Michigan, one billion birds were supposedly killed. By the late 1800s, less than one million Passenger Pigeons were left and most of these were members of one flock. Hunters found out where the flock would be by telegraph and searched the birds out and killed them. In 1914, the last Passenger Pigeon—a bird born in captivity in the Cincinnati Zoo—died.

Insular species, many often endemic, are usually more easily extirpated than mainland species primarily because they have nowhere to go for escape. Today, on Mauritius, where the Dodo was found, 300 of the 1100 plant species known on the island, 28% of the total flora, are presently threatened. Another group of island residents there who have suffered from the presence of man are the giant tortoises of the genus *Geochelone*. These tortoises were known only from the Galapagos Islands in the eastern Pacific and the Mascarenes (the island chain of which

Mauritius is a member) and the Seychelles in the Indian Ocean. By the early 1800s, all but one of the Indian Ocean tortoises had been extirpated and four species of the Galapagean turtles were gone (and most of the remaining 10 species threatened). By a strange quirk of fate, the journal of a French Huguenot resident on Mauritius, published in 1708, may have sped up the demise of the Mascarene turtles. There were groups of as many as 2,000-3,000 tortoises on the island and their meat was excellent food, he pointed out in his journal. Unfortunately, both the French and English navys discovered this easily accessible larder of meat—the Mascarenes were on the Indian Ocean's main shipping routes, and the Mascarene Islands giant tortoises were extinct by the late 1700s.

Habitat Destruction, Threats to Humans

Plants and animals that require unusual habitats, have large home ranges, and appear to be threats—real or imagined—to humans are, as history has proven, at risk to be extirpated by humans and their activities. The Ivory-billed Woodpecker (*Campephilus principalis*), though never common, was known from the old-growth bottomlands of the Atlantic and Gulf Coastal Plains of the southeastern United States. The Ivorybill was found only in large tracts of old-growth bottomlands, evidently requiring a sizeable home range. The bird preferred to feed on beetle grubs found in the rotting boles of naturally-decaying hardwood species. From the late 1800s into the 1900s, forestry and agricultural operations dramatically reduced the size and frequency of old-growth tracts in the Southeast. By the 1950s, only a few old-growth bottomlands of any size were left.

The Ivorybill was a large, spectacular woodpecker with an ivory-white bill. Its beauty may have also contributed to its demise. Mark Catesby, an early naturalist-explorer of the Carolinas, pointed out that:

> The bills of these Birds are much valued by the *Canada Indians*, who made Coronets of 'em for their Princes and great warriors, by fixing them round a Wreath, with their points outward. The Northern Indians, having none of these Birds in their cold country, purchase them of the *Southern People* at the price of two, and sometimes three, Buck-skins a Bill.

Ivorybill skins were also highly-valued by ornithologists in the late 1800s, when most collectors already knew that this was a rare bird. Arthur T. Wayne, an ornithologist who couldn't find any birds near his home in Charleston, South Carolina, supposedly saw around 200 Ivorybills in Florida in 1892, 1893, and 1894. In 1892 alone, he killed 13 birds for their skins; the skins yielded $20 each from museums.

The Ivory-billed Woodpecker is still part of the myth and lore of large, southeastern floodplain forests in the Southeast. Although not one bird has been documented in North America in the last 40-50 years, the species is the subject of books and papers and has been the central focus of several bottomland conservation campaigns. As recently as 2001, an expedition of ornithologists was sent out to Louisiana to search for the bird. It is thought that there are a few birds still alive in remote portions of eastern Cuba, though the population size of the Ivorybills there is not known.

Predators such as large cats, bears, wolves, and crocodilians can potentially devour the smaller, slower *Homo sapiens* and have always be seen as a threat to man. Historical literature and anecdotes have grossly exaggerated the actual threat these larger creature pose; however, the large predatory mammals have to live with their bad press.

The tiger (*Panthera tigris*) is one of the most fascinating creatures that lives or has ever lived. At one time, tigers were found from the Black Sea to the island of Bali in Indonesia. There were originally eight subspecies of the tiger, but today only five of these subspecies are extant and three of the remaining five subspecies are endangered. Only the Bengal Tiger (*Panthera tigris tigris*) and the Indochinese Tiger (*Panthera tigris corbetti*) have populations with over 1000 individuals (Table 4). The Balinese Tiger, the Caspian Tiger, and the Javan Tiger all were extirpated in the 1900s, and the Siberian Tiger, the Amoy Tiger, and the Sumatran Tiger, all with fewer than 500 individuals known, could become extinct in our lifetime. Because of the tiger's ferocity as a predator, it may be impossible for man and the tiger to live in close proximity. One day soon, all known tigers will be in remote nature reserves, unpopulated areas, and zoos.

Fewer tears are shed for vanishing plants than animals. The logo of the World Wildlife Fund, now the panda (*Ailuropoda melanoleuca*), will probably never be changed to *Rafflesia* sp., a rare parasitic plant with the world's largest flower (a large purple, foul-smelling giant). Because plants do not move, they are the organisms more closely associated with place or habitat. A tropical rainforest is a rainforest because of the density of plants, not animals, there.

For thousands of years, ginseng (from the Chinese "man root," referring to the shape of its root), has been a part Chinese herbal medicine. Chinese ginseng and Korean ginseng have been

Table 4. World status of the tiger (*Panthera tigris*).

SUB-SPECIES	COMMON NAME	DISTRIBUTION	EST #
P. tigris altaica	Siberian Tiger	China, Korea, Russia	350-400
P. tigris balica	Balinese Tiger	Bali	Extirpated 1940s
P. tigris tigris	Bengal Tiger	Bangladesh, Bhutan, China, India, Myanmar, Nepal	3100-4500
P. tigris virgata	Caspian Tiger	Afghanistan, Iran, Russia, China, Turkey	Extirpated 1970s
P. tigris corbetti	Indochinese Tiger	Cambodia, China, Laos, Malaysia, Myanmar, Thailand, Vietnam	1200-1700
P. tigris sondaica	Javan Tiger	Java	Extirpated 1980s
P. tigris amoyensis	South China Tiger	China	20-30
P. tigris sumatrae	Sumatran Tiger	Sumatra	400-500

Data Source: Edited from table at www.5tigers.org, May 1998.

collected extensively for centuries. Chinese and Asians all over the world have used ginseng as an aphrodisiac, energy booster, appetite enhancer, and so on. By the late 1800s, ginseng was extremely rare in the wild in China. The price per pound had risen dramatically in Hong Kong and Taiwan, where the Asian herbal trade is centered.

By historical accident, the flora of eastern North America is similar to that of Japan and eastern China (see Chapter IV). Asian eyes, therefore, turned to American ginseng (*Panax quinquefolius*), which is found in rich, deciduous woods from Canada to northern Florida. By the 1930s and 1940s, collectors were combing every ravine and valley in the southern Appalachians looking for "sang," as it is locally called. Collecting took place in the fall when the berries could be spotted. Plants would be pulled up (some conscientious collectors would spread the berries that unbeknownst to them take about 10 years to grow into a mature plant) and hauled away to dealers. In the 1970s and 1980s, when conservation agencies began to inventory eastern states for rare plant species, it was discovered that many ravines and valleys from which ginseng had formerly been collected no longer harbored the species. Exploitation for the Asian market had nearly driven the species to extinction is some states. Collection of ginseng is now illegal or closely regulated in most eastern states.

Habitats and Species

Most western governments now go to unprecedented lengths to protect endangered and threatened plants and animals in their countries, and international organizations such as the World Wildlife Fund, the International Union for the Conservation of Nature, and The Nature Conservancy are now actively working in third world countries. Zoological gardens in the West spend enormous amounts of money to "protect" animals, as well as to secure individuals of rare species for display and captive breeding.

As most conservation organizations have already realized, it is very late in the day to do things piecemeal. Human population is growing at geometric rates, and natural resources and the natural world itself is disappearing faster each day. It is evident that we need to protect vast expanses of land on all of the continents. The World Heritage List published by UNESCO has documented around 150 premier world natural areas from 122 countries. These areas include the best natural parks and natural areas of most countries of the world. These sites, however, are just the tip of the iceberg. Always an adage, but so true, we must think globally and act locally. That virgin forest in your state, that wetland in your county, that forest in your backyard..., you must try to protect them all.

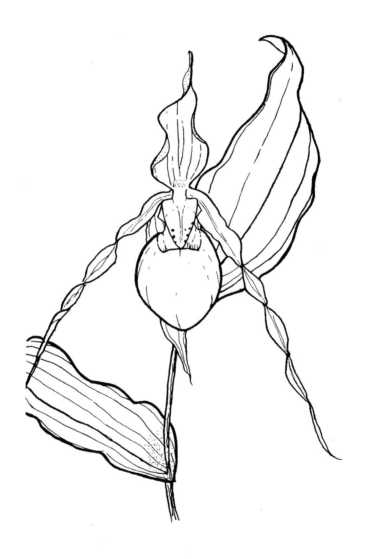

Figure 20. The yellow lady's-slipper (*Cypripedium pubescens*), an orchid.

X

WHAT A PIECE OF WORK IS MAN?

Still thou are blest, compared wi' me!
The present only toucheth thee:
But och! I backward cast my ee
On prospects drear!
And forward, though I cannot see,
I guess and fear.

Robert Burns, "To a Mouse"

Therefore, since the world has still
Much good, but much less good than ill,
And while the sun and moon endure
Luck's a chance but trouble's sure,
I'd face it as a wise man would,
And train for ill and not for good

A.E. Housman, "Terence, This is Stupid Stuff"

To his horror he [Toad] recollected that he had left both
coat and waistcoast behind him in the cell, and with
them his pocket-book, money, keys, watch, matches,
pencil case—all that makes life worth living, all the
distinguishes the many-pocketed animal, the lord of
creation, from the inferior one-pocketed or no-pocketed
productions that hop or trip about permissively,
unequipped for the real contest.

Kenneth Grahame, The Wind in the Willows

Origins

Man or something fairly similar to us has been around for about 2.5 million years. Most anthropologists believe than *Homo sapiens* arose in Africa and moved across the face of the globe from there. The species *Homo sapiens* is often given as an example of the "founder principle," by which a species arises from a small population in one place and spreads to a larger range. Such species are not known for their genetic diversity. Although we now inhabit every continent of the earth, the "genetic" distance from individual to individual of different "races" of our species is not very far. Many attempts have been made to prove that the human races are physically different from one another, but no scientific study has proven that these differences are anything but superficial. The primary differences between people of different races appear to be cultural. An African-born individual raised in London will have little cultural connection to a similarly-born child brought up in Peoria, Illinois or Dar Es Salaam, Tanzania. Skin color is a superficiality in human races; culture is what matters.

Man, the Measure of All Things

Philosophers have posed many questions concerning the nature of man the beast versus man the vessel of the "soul." The question concerning man that generates the greatest fascination in the biologist and the naturalist is: "Is man really different from other animals?" It leads on to other similar basic questions such as: "How is man different from other creatures?" and "Is man unique?"

It appears that man has about 30,000 plus genes, about the same as the chimpanzee, and about 11,000 more than a rat. As what Jared Diamond calls the "third chimpanzee," we are only different from the other two species of chimps in 1.8% of our genetic make-up. Compare this to the Red-eyed Vireo (*Vireo olivaceous*) and the White-eyed Vireo (*Vireo griseus*), congeners who are 3% different genetically. What, therefore, separates man from chimpanzee?

There are some things we do that no other animal has ever done. There are over 6000 languages on the face of the earth, each one with many nuances of thought absent in the others. Chimps can be taught to say a few words and perform basic communication skills in a few languages, but with our languages man can not only communicate but can weave fanciful stories, chant poetry, and sing and play music unknown to any other animal (as far as we can understand animal communication). Language, song, and music are, at the least, *among* the major attributes that separate man from beast. I once had a professor of literature who used to do a wonderful impersonation of God receiving mankind on judgment day. God asks, "What have you done?" "What have you to show me?" Man shows his achievements; God laughs at every symbol of man's greatness. Finally, the sound of Beethoven's Ninth Symphony is heard in the distance, God falls on his knees and begins to weep with joy, so the story goes.

Secondly, man, unlike beast, can observe the world around him and make analytical judgments of what he sees, including himself in his purview of nature. This ability, which the late anthropologist Loren Eiseley called "the eye in the skull," is what makes you and me interested in other creatures, aside for food or territorial reasons. This type of consciousness may have arisen over the last few thousand years, some psychologists point out.

Man's analytical ability allows him to live in the present, past, and future, unlike Mr. Burns' field mouse, who only knows the present.

And, thirdly, there is clothing. In the Judaeo-Christian bible in the book of Genesis, after God found that Adam and Eve had eaten of the fruit of the tree of knowledge, he clothed the two with "coats of skins" and drove them out of the Garden of Eden. In *The Wind in the Willows*, when the noble and wealthy Toad (Figure 21) escapes from jail, he departs in such a hurry that he leaves his coat and waistcoast (vest to Americans) behind.   Subsequently, Kenneth Grahame, the author, satirizes Toad: "...with them [coat and waistcoat...he had left] his pocketbook, money, keys, watch, pencil case—all that makes life worth living, all that distinguishes the many-pocketed animal, the lord of creation, from the inferior one-pocketed or no-pocketed productions that merely hop or trip about permissively, unequipped for the real contest. "

Man is, of course, a "many-pocketed" animal, the zenith of evolution and fashion. And we are always staring into mirrors to make sure our clothes are right. Our clothes conceal from us and from others the beauty or the beast inherent in our nakedness. Clothing helps us cover the fact that, although our brains are far advanced over other animals (in most cases), our bodies still reveal that we are animals.   They cover the delicate, sensuous breasts of women (the sight of which turn men into animals), and the hairy backs and chests of men that show men are really animals.

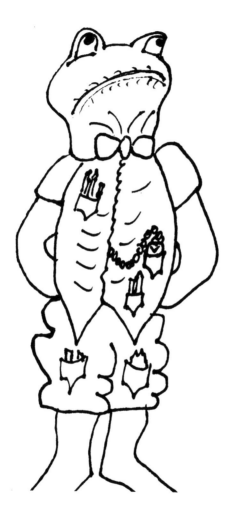

Figure 21. Toad from *The Wind in the Willows*.

When did man the animal become man the Man? What year, what day? Was the change sudden? Are we descendants of colonists from another universe? Or are we the logical end of a strange evolutionary line of weird primates? Over the last few decades while biologists were decoding and mapping the human DNA, climatologists and geologists (with much less fanfare) have been drilling through the Greenland ice cap. Sorting through what some have called the earth's "ice memory," they have discovered that the earth's climate was not a gentle one in the past, but included dramatic and sudden alterations and extremes. Researchers in Greenland cannot find a recent period of relatively stable climate similar to the one in which we are currently living. Some scientists think that man, the roaming hunter-gatherer, was able to settle down and grow crops (several anthropologists assert that the only reason we settled down anyway was to grow grains for beer and grapes for wine) and become "civilized" because climate has remained relatively gentle since the late Pleistocene. Civilized, stable agricultural man as we know him has only been around for 7000 to 8000 years, dating back to early Chinese civilization. Worst case scenario, we therefore owe the rise of civilization to good weather and our love for alcohol—surf's up, dudes, enjoy.

We, it seems, are presently living in a golden age without major cataclysmic climatic or geological events (see Chapter III for a discussion of mass extinctions caused by extraterrestrial bodies striking the earth in). In other words, we may owe everything we have accomplished to this brief shining moment of exceptionally good weather on the face of our green planet. The rise of civilization, music, art, and the birth of technology have all occurred in the last few seconds of the hour in the earth's scale of time. Some have asserted that the rise of the bicameral mind and of man's consciousness of himself is the evolutionary event that has separated man from beast and, subsequently, paved the way to civilization. All of these events, they postulate, needed ideal conditions to develop and only occurred in recent times. If this is true, we must heed the words of A. E. Housman, "Luck's a chance, but trouble's sure,/ I'd face it as a wise man would,/ And train for ill and not for good. "

Man, the Master of All Things

> [And Man shall] have dominion over the fish of the sea, and over the fowl of the air, and over the cattle, and over all of the earth, and every creeping thing that creepeth upon the earth.
>
> Genesis 1:26

Man, more so than any other earthly creature, dramatically alters his habitat. Unlike his simian relatives, he tears down the forest to build his house out of the wood of the trees. He digs into the earth, searching for elements to use to create massive structures, such as the majestic Great Wall of China, purportedly the only ancient man-made object visible from space. Like the Great Wall, the cathedral at Chartres, and the city of Florence, some of the edifices man erects are, without question, art; but some of the creations of man—the mobile home, industrial districts of third world countries and of anywhere U. S. A.—fall into other categories.

Today, one can fly over the Amazon basin or the North American Pacific Northwest and spot large, deforested landscapes resulting from the clearcutting of old-growth trees over 200 years old. In colonial times, it would have taken a group of men with axes years to do what now can be done in a few weeks or months. As technology progresses, man's ability to manipulate the landscape becomes easier. We have diverted—and rediverted (when we discovered we had made a mistake by diverting)—rivers; impounded rivers and creeks, flooding vast areas of land; cleared much of the earth for agriculture; seeded clouds; split atoms; cloned sheep; and we have created and/or propagated deadly pathogens such as Legionnaire's disease and the HIV virus.

When one surveys the natural landscape in and around populated areas, one of the first things noticed is the high proportion of exotic species or introductions present in any local fauna or flora in the developed world. First of all, nearly all of our crops, which have evolved with man, and their incumbent weeds, which have coevolved with the crops, are introduced species. Secondly, our livestock, our pets, our lawns, our gardens (vegetable or flower) are predominantly species that have originated or have been developed in countries other than our own. In the past, we became fascinated with exotic species, willingly introducing them into an alien environment (or they have introduced themselves to us, as Michael Pollan puts it in the *Botany of Desire*). Many of these species prospered without competition and became pests in their new habitat. More recently, globetrotting man has unwillingly moved species from continent to continent, creating massive ecological problems.

The Future

> We have already appropriated 40 percent of the planet's organic matter produced by green plants. If everyone agreed to become vegetarian, [the earth] could support about 10 billion people. If humans utilized as food all of the energy captured by photosynthesis on land and see, some 40 trillion watts, the planet could support about 16 billion people.
>
> "The Future of Life," E. O. Wilson

The epilogue of most works on conservation biology and biodiversity usually consists of statements about vanishing rainforest and recently-extirpated species, along with warnings of imminent doom. Here, I want to take a more positive, philosophical approach. Man is really the strangest of all animals. We can kill; we can destroy. But our ultimate potential is still undiscovered. With "the eye in the skull," there is so much for that eye to see, and the skull houses an amazing organ for recording and understanding what the eye sees. I am writing this text on a computer, an imitator of the actions of the human brain, a machine designed by man. Our historians of philosophy convince us that what really matters is our "rational" mind. If we are to survive on this planet, reason must indeed prevail over international differences and emotional religious disputes. Reason, however, is not everything. We need an inner force to sustain through the some of the monotonous and Sisyphean tasks of life. Every one of us intuitively knows that without creativity, imagination, and laughter, there would be no need to continue living. As said by no less an intellect than Albert Einstein: "The intuitive mind is a sacred gift and the rational mind is a faithful servant. We have created a society that honors the servant and has forgotten the gift."

What is there left for us to do? Though we are territorial animals, we must, as Yeats said, "teach the free man how to praise," instead of teaching how to fight (we already know how to do that). We must continue to see and hear the sacred and the beautiful in nature. The man in the Bob Dylan's song who was not "busy being born" was "busy dying." In my introduction, I said that the diversity of life is an epic poem written in an international language understood by all who live in the natural world. I think that nature, our mother, is one of the keys to uniting the world. With our understanding of nature, with a sense of wonder, a sense of the holy, and a sense of what is moral, we can better understand and take care of the earth, its creatures, and its people. And we must, because it is all we have and all we will ever have.

SOURCES AND FURTHER READINGS

CHAPTER I

Barnes, R. D. 1968. Invertebrate zoology. W. B. Saunders, Philadelphia.

Ellis, R. 1998. The search for the giant squid. Penguin, New York.

Margulis, L. and K. V. Schwartz. 1998. Five kingdoms: an illustrated guide to the phyla on Earth. W. H. Freeman, New York.

Pennak, R. W. 1978. Freshwater invertebrates. John Wiley, New York.

Tudge, C. 2000. The variety of life on earth: a survey and a celebration of all the creatures that have ever lived. Oxford University Press, Oxford.

CHAPTER II

Darwin, C. 1839. Journal of researches into the geology and natural history of the various countries visited by H.M.S. Beagle under the command of Captain Fitzroy, R. N. from 1832 to 1836. Henry Colburn, London.

Gaston, K. J. 1996. Biodiversity: a biology of numbers and difference. Blackwell Science, Oxford.

_____ and J. I. Spicer. 1998. Biodiversity: an introduction. Blackwell Science, Oxford.

Holldobler, B. and E. O. Wilson. 1990. The ants. Harvard, Cambridge, MA.

May, R. M. 1988. How many species are there on earth? Science 241: 1441-1449.

Wallace, A. R. 1869. The Malay Archipelago: the land of the orangutan and the bird of paradise (Facsimile edition). Harper, New York.

Wilson, E. O. 1975. Sociobiology: the new synthesis. Harvard, Cambridge.

Zimmerman, E. C. 1958. Three hundred species of Drosophila in Hawaii? A challenge to geneticists and evolutionists. Evolution 12:557-558.

CHAPTER III

Alvarez, L. W., W. Alvarez, F. Asaro, and H. V. Michel. 1980. Extraterrestrial cause for the Cretaceous-Tertiary extinction. Science 201:1095-1108.

Behrensmeyer, A. K., J. D. Damuth, W. A. DiMichele, R. Potts, H. D. Sues, and S. L. Wing. Editors. 1992. Terrestrial ecosystems through time: evolutionary paleoecology of terrestrial plants and animals. Chicago, University of Chicago Press.

Darlington, P. J. 1975. Biogeography of the southern end of the world: distribution and history of far southern life and land, with an assessment of continental drift. Cambridge, Mass., Harvard University Press.

Hallam, A. and P. B. Wignall. 1997. Mass extinctions and their aftermath. Oxford University Press, New York.

Janzen, D. H. and P. S. Martin. 1982. Neotropical anachronisms: the fruit the gomphotheres ate. Science 215: 19-27.

Raup, H. M. 1991. Extinction. Oxford, Oxford University Press.

Richards, M. A., R. G. Gordon, and R. D. van der Hilst. Editors. 2000. The history and dynamics of global plate motions. Geophysical Monograph 121. American Geophysical Union, Washington.

Ross, C. A. Editor. 1976. Paleobiogeography. Benchmark Papers in Geology/31. Dowden, Hutchinson & Ross, Shroudsburg, PA.

Tallis, J. H. 1991. Plant community history: longterm changes in plant distribution and diversity. Chapman and Hall. London.

Wegener, A. 1966. The origins of continents and oceans. Translation of 1929 edition of 1915 work from German by J. Biram. Dover, New York.

CHAPTER IV

Brown, J. H. and A. C. Gibson. 1983. Biogeography. C. V. Mosby, St. Louis.

Browne, J. 1983. The Secular Ark: studies in the history of biogeography. Yale University Press, New Haven.

MacArthur, R. H. and E. O. Wilson. 1967. The theory of island biogeography. Monographs in population biology, No. 1. Princeton University Press, Princeton, N. J.

Sclater, P. L. 1858. On the general geographical distribution of the members of the class Aves. Zoological Journal of the Linnean Society 2:130-145.

Taktajian, A. 1978. The floristic regions of the world (translated from the Russian). Nauka, Leningrad.

Urquhart, F. A. 1960. The monarch butterfly. University of Toronto Press, Toronto.

Wallace, A. R. 1876. The geographical distribution of animals. 2 vol. Macmillan. London.

CHAPTER V

Antonovics, J. 1968. Evolution in closely adjacent populations. V. Evolution of self-fertility. Heredity 23:219-238.

_____ and A. D. Bradshaw. 1970. Evolution in closely adjacent populations. VII. Clinal patterns at a mine boundary. Heredity 25:349-362.

Briggs, D. and S. M. Walters. 1997. Plant variation and evolution. Cambridge (U.K.), Cambridge University Press.

Darwin, C. 1859. On the origin of species by means of natural selection, or the preservation of favoured races in the struggle for life. John Murray, London.

Eldridge, N. Editor. 1987. The *Natural History* Reader in Evolution. Columbia University Press, New York.

Fryer, G. and T. D. Iles. 1972. The cichlid fish of the Great Lakes of Africa: their biology and evolution. Edinburgh, Oliver & Boyd.

Gould, S. J. 2002. The structure of evolutionary theory. Harvard, Cambridge.

Pollan, M. 2001. The botany of desire: a plant's-eye view of the world. Random House, New York.

CHAPTER VI

Harper, J. L. 1977. The population biology of plants. Academic Press, London.

Janzen, D. H. 1976. Why bamboos wait so long to flower? Annual Review of Ecology and Systematics 7:347-391.

Lewontin, R. C. 1968. Population biology and evolution. Syracuce University Press, Syracuse, NY.

CHAPTER VII

Majecki, J. 1983. [*Beraeodes minutus* on *Limnephilus rhombicus*] Acta Univ Lodziensis Folia Limnol 1: 51-60.

Barnes, R. D. 1968. Invertebrate zoology. W. B. Saunders. Philadelphia.

Beattie, A. J. and P. R Erlich. 2001. Wild solutions. Yale University Press, New Haven.

Colinvaux, P. A. 1978. Why big fierce animals are rare. Princeton University Press, Princeton, N. J.

Emerton, J. H. 1961. The common spiders of the United States. Dover Publications, New York (republication of 1902 edition).

Faegri, K, and L. van der Pijl. 1979. The principles of pollination biology. Pergamon Press, Oxford.

Forsythe, A. 1986. The natural history of sex. Chapters, Shelburne, Vermont.

Gaddy, L. L. and J. H. Morse. 1985. Common spiders of South Carolina. Agricultural Experiment Station Bulletin. Clcmson University, Clemson, S. C.

Gause, G. F. 1934. The struggle for existence. The Williams and Wilkins Co., Baltimore, MD.

Ghiselin, M. T. 1974. The economy of nature and the evolution of sex. University of California Press, Berkeley.

Hutchinson, G. E. 1959. Homage to Santa Rosalia, or Why there are so many kinds of animals? American Naturalist 93:145-159.

Janzen, D. H. 1977. What are dandelions and aphids? American Naturalist 111: 586-589.

CHAPTER VIII

Borman, F. H. and G. Likens. 1979. Pattern and process in a forested ecosystem: disturbance, development, and the steady-state base—Hubbard Brook ecosystem study. Springer-Verlag, New York.

Cody, M. L. and J. M. Diamond. Editors. 1975. Ecology and evolution of communities. Harvard (Belknap Press), Cambridge.

Jacobson, M. C., R. J. Carlson, H. Rodhe, and G. H. Orians. editors. 2000. Earth system science: from biogeochemical cycles to global change. International Geophysics Series, Vol. 72. Academic Press, New York.

Whittaker, R. H. 1975. Communities and ecosystems. Macmillan, New York.

CHAPTER IX

Catesby, M. 1746. Natural history of the Carolinas, Florida, and the Bahamas. Privately published in London.

Day, D. 1981. The doomsday book of animals: a natural history of vanished species. Viking Press, New York.

Eldredge, N. 1991. The miner's canary. Prentice-Hall Press, New York.

Flannery, T. and P. Schouten. 2001. A gap in nature: discovering the world's extinct animals. Atlantic Monthly Press, Boston.

Quammen, D. 1996. Song of the Dodo: island biogeography in an age of extinctions. Simon & Schuster, New York.

CHAPTER X

Diamond, J. M. 1992. The third chimpanzee. HarperCollins. New York.

_____. 1997. Guns, germs, and steel. W. W. Norton. New York.

Eiseley, L. C. 1967. The immense journey. Vintage Books, New York.

_____. 1978. The star thrower. Harcourt Brace Jovanovich, New York.

Grahame, Kenneth. 1940. The wind in the willows. Heritage Press. New York.

Lamb, H. H. 1982. Climate history and the modern world. Methuen, London.

Marks, J. M. 1995. Human biodiversity. Aldine de Gruyter. New York.

INDEX